Buildings for Advanced Technology

Science Policy Reports

The series Science Policy Reports presents the endorsed results of important studies in basic and applied areas of science and technology. They include, to give just a few examples: panel reports exploring the practical and economic feasibility of a new technology; R & D studies of development opportunities for particular materials, devices or other inventions; reports by responsible bodies on technology standardization in developing branches of industry.

Sponsored typically by large organizations – government agencies, watchdogs, funding bodies, standards institutes, international consortia – the studies selected for Science Policy Reports will disseminate carefully compiled information, detailed data and in-depth analysis to a wide audience. They will bring out implications of scientific discoveries and technologies in societal, cultural, environmental, political and/or commercial contexts and will enable interested parties to take advantage of new opportunities and exploit on-going development processes to the full.

More information about this series at http://www.springer.com/series/8882

Ahmad Soueid • E. Clayton Teague •
James Murday

Editors

Buildings for Advanced Technology

 Springer

Editors
Ahmad Soueid
HDR Architecture, Inc.
Alexandria, Virginia
USA

E. Clayton Teague
Gaithersburg, Maryland
USA

James Murday
Office of Research Advancement
Univ of Southern California
Washington, District of Columbia
USA

ISSN 2213-1965
Science Policy Reports
ISBN 978-3-319-79694-9
DOI 10.1007/978-3-319-24892-9

ISSN 2213-1973   (electronic)

ISBN 978-3-319-24892-9   (eBook)

Cover illustration: Photo courtesy of HDR Architecture, Inc.; © Steve Hall © Hedrich Blessing

Printed on acid-free paper

Springer International Publishing AG Switzerland is part of Springer Science+Business Media (www.springer.com)

Preface

The principal editors have each been engaged in the design and construction of buildings for nanoscale science and engineering research and have noted the absence of an overarching document to guide such activity. In seeking to develop the necessary information and data that would be useful in designing and constructing buildings for such advanced technologies as nanotechnology and biotechnology, the editors convened three workshops at which architects, building contractors, environmental control specialists, and scientists engaged in these technologies were contributors. Over 300 participants attended these workshops convened from 2003 to 2006. Specialists and experts with knowledge and experience in the control of environmental disturbances to buildings and experimental apparatus contained therein contributed to these workshops and to this document. The document compiles digests of inputs from the workshops' participants with the expertise of other selected specialists and user scientists.

The design and engineering challenges identified at the workshops for nanotechnology facilities include:

- Establishing and maintaining critical environments: temperature, humidity, and pressure
- Structural vibration isolation
- Airborne vibration isolation (acoustic noise)
- Isolation of mechanical equipment-generated vibration/acoustic noise
- Cost-effective power conditioning
- Grounding facilities for low electrical interference
- Electromagnetic interference (EMI)/Radio frequency interference (RFI) isolation
- Airborne particulate contamination
- Airborne organic and chemical contamination
- Environment, safety and health (ESH) considerations
- Flexibility strategies for nanotechnology facilities

The document addresses the technology challenges unique to each of these areas and provides best practices and examples of engineering approaches to address these challenges. It presents the ideas that were generated by the various breakout groups at the workshops and supplemented by contributions from professional building design engineers and architects. The objective of the document is to provide insights on the design priorities and trade-offs for buildings to be utilized for nanotechnology, and other environmentally sensitive, research. From the time of the workshops, some changes in standards and approaches to designing and constructing buildings have occurred. Therefore, the editors have updated the recommendations, best practices, and standards to be current as of 2014.

Without the major contributions from the authors thanked in the Acknowledgements section, as well as all the other contributors and workshop participants listed in Appendix B, this identification of the challenges and compilation of information and data to guide the design and construction of buildings for advanced technologies would not have been possible. The editors extend our deep gratitude to all the contributors.

Alexandria, VA	Ahmad Soueid
Gaithersburg, MD	E. Clayton Teague
Washington, DC	James Murday

1. Brookhaven National Laboratory, Center for Functional Nanomaterials (CFN), Upton, NY, Characterization lab, "Photo courtesy of HDR Architecture, Inc.; ©2008 Mark Boisclair"

2. Argonne National Laboratory, Energy Sciences Building, Lemont, IL, Entrance façade, "Photo courtesy of HDR Architecture, Inc.; ©2013 Dave Burk/Hedrich Blessing"

3. Ahmad Soueid, "Photo courtesy of HDR Architecture, Inc."

4. Argonne National Laboratory, Energy Sciences Building, Lemont, IL, Ultra-High Vacuum (UHV) in characterization lab, "Photo courtesy of HDR Architecture, Inc.; ©2013 Dave Burk/Hedrich Blessing"

5. Gateway University Research Park, Joint School of Nanoscience and Nanoengineering (JSNN), Greensboro, NC, Specialty lab, "Photo courtesy of HDR Architecture, Inc.; ©2013 Mark Herboth"

6. Sandia National Laboratories, Center for Integrated Nanotechnologies (CINT) Core Facility, Albuquerque, NM, Molecular Beam Epitaxy (MBE) Lab, "Photo Courtesy of HDR Architecture, Inc.; ©2007 Mark Boisclair"

7. Purdue University, Birck Nanotechnology Center, West Lafayette, IN, Main entrance, "Photo courtesy of HDR Architecture, Inc.; Steve Hall ©Hedrich Blessing 2008"

8. University of Arkansas, Nanoscale Materials Science and Engineering Building, Fayetteville, AK, Entrance façade, "Photo courtesy of HDR Architecture, Inc.; ©2011 Dero Sanford"

9. National Institute of Standards and Technology (NIST), Advanced Measurement Laboratory (AML), Gaithersburg, MD, Advanced Technology Lab, "Photo courtesy of HDR Architecture, Inc.; ©Steve Hall ©Hedrich Blessing"

10. Brookhaven National Laboratory, National Synchrotron Light Source II (NSLS-II), Upton, NY, Magnets in storage tunnel, "Photo courtesy of HDR Architecture, Inc.; ©2014 David Sundberg/ESTO"

11. National Institute of Standards and Technology (NIST), Precision Measurement Lab (PML), Boulder, CO, Cleanroom, "Photo courtesy of HDR Architecture, Inc.; ©2012 Andrew Pogue Photography"

12. Brookhaven National Laboratory, National Synchrotron Light Source II (NSLS-II), Upton, NY, Aerial view, "Photo courtesy of HDR Architecture, Inc.; ©2014 David Sundberg/ESTO"

Acknowledgments

E. Clayton Teague, Guest Researcher at NIST (and former Director of the U.S. National Nanotechnology Coordination Office—NNCO), James S. Murday, University of Southern California (and formerly of the Office of Naval Research and former NNCO Director), and Ahmad Soueid of HDR Architecture, Inc., as compilers and principal authors of this report wish to thank other authors whose contributions added significant depth to this report[1]:

Terry Abair
SUNDT Construction, Inc.
Larry Allard
Oak Ridge National Laboratory
Hal Amick
Colin Gordon & Associates
Brett Dominguez
DPR Construction, Inc.
Michael Gendreau
Colin Gordon & Associates
David Gibney
HDR Architecture, Inc.
Phil Haswell
University of Alberta
Brett Helm
DPR Construction, Inc.
Thomas C. Isabell
JEOL USA, Inc.

Mark Jamison
HDR Architecture, Inc.
Tim Loughran
AdvanceTEC, LLC
Ralph Morrison
Consultant
Greg Parker
Currie & Brown, Inc.
Neal Shinn
Sandia National Laboratories
Michael Somin
Earl Walls Associates
Mark Stephens
EPRI PEAC Corporation
Chuck Thomas
EPRI PEAC Corporation
Norm Toussaint
HDR Architecture, Inc.

---

[1] Affiliations are as of the time of the workshops.

Eric Ungar
Acentech, Inc.
Lou Vitale
VitaTech Engineering

John Weaver
Purdue University
Amir Yazdanniyaz
ARUP
Ted Zsirai
HDR Architecture, Inc.

The information in this document is derived in part from a series of workshops which were jointly sponsored by the National Institute of Standards and Technology, the Office of Naval Research, and the U.S. Department of Energy. Additional sponsorship was provided by Arizona State University, Purdue University, and HDR Architecture, Inc. We wish to thank all the participants at each of these three Buildings for Advanced Technology (BAT) workshops, held January 14–16, 2003, at the National Institute of Standards and Technology in Gaithersburg, MD; January 21–23, 2004, at Arizona State University, Mesa, AZ; and February 7–8, 2006, at Purdue University in West Lafayette, IN. The presentations and discussions at these workshops, supplemented by input from building design professionals and more recent literature, make up much of the content of this document. Special thanks to those experts who contributed much of the incorporated information.

Special thanks to Geoff Holdridge, Ellen Randall, Cheryl David-Fordyce, and Pat Johnson for their editing of the report.

Contents

Chapter 1
Introduction

Background

Nanotechnology is the understanding and control of matter at dimensions between approximately 1 and 100 nm. Encompassing nanometer-scale science, engineering, and technology, nanotechnology involves imaging, measuring, modeling, and manipulating matter at this length scale. Nanostructured materials can exhibit unusual physical, chemical, and biological properties and can enable novel applications not possible in bulk materials of the same chemical composition.

The use of nanostructures in technology has been around for centuries, so why is it now all the rage? The capability to extensively measure and manipulate individual nanostructures has only been available from the 1990s, with the development of new highly engineered enhanced versions of analytical tools. These include advances in scanning probe microscopies/spectroscopies as well as in high-resolution electron microscopy. Properties of an individual nanostructure, rather than ensemble averaged property values, can now be observed and fully exploited. In turn, those properties can be understood in terms of structure and composition. With this understanding comes the possibility to tailor and reliably reproduce these properties and to accelerate progress towards new technologies.

The economic engines driving nanoscale research include industries most likely to benefit, such as information technology, biotechnology, renewable energy, and high-performance materials. Estimates of the economic impact of nanotechnology cite a worldwide commercial market on the order of three trillion dollars per year by 2020 for products whose function is enabled by the properties of nanostructures (i.e., products that are "nano inside"). The nanometer size is noteworthy for one additional reason—it is the last size scale to be addressed by materials science. Chemistry and atomic/molecular physics have delved deeply into smaller structures; solid-state physics and materials science into scales at micrometer and above.

© Springer International Publishing Switzerland 2015
A. Soueid et al. (eds.), *Buildings for Advanced Technology*, Science Policy Reports,
DOI 10.1007/978-3-319-24892-9_1

The new understanding of nanoscale structures fills the gap and holds promise for an integration of simulation/modeling efforts into a coherent, truly predictive approach to materials behavior.

Surface science can be considered one-dimensional nanoscience. The number of atoms available for measurement is $\sim 10^7$ (surface layer bounded inside a square micrometer). Further, surface science needs a special environment provided by ultra-high-vacuum (UHV) systems both to protect the carefully prepared surfaces and to provide a meter-scale mean free path for the photon/electron/atom/ion probes. However, other than reasonable cleanliness to avoid contamination, the requirements imposed on building environments are not severe.

Nanoscience has provided tools capable of single-atom measurements and sub-angstrom spatial resolution. But an atom does not provide much signal. To maximize the signal-to-noise, analytical tools at the nanoscale tend to use sharp tips or focused beams positioned on the atom-sized scale. Centimeter- to meter-scale analytical tools are counted on to deliver precise, stable positioning at the sub-nanometer scale. Environmentally induced distortions on tools of meter size scales easily perturb measurements at the sub-nanometer size scale resolution. So a wide range of variables must be controlled, such as ambient air pressure differentials between building spaces, acoustic (air) and structurally propagated vibrations, temperature and humidity, fluctuations in electrical power, electromagnetic interferences, and molecular/particulate contamination. Equipment engineering can and does reduce those distortions but does not eliminate them. Further, the development of new analytical tools generally does not yet have such engineering incorporated, so control of the building environment becomes even more critical.

To this challenge, one must add another, molecular biology, which is unveiling the mysteries of cellular physiology. If one considers the cell as a factory of micrometer dimensions, it is clear that the many molecular machines inside that factory are nanometer in scale. Nanoscience is certainly accelerating the rate of progress in molecular biology. In addition, as semiconductor devices have shrunk to the nanoscale, there is growing opportunity to more effectively couple biological and semiconductor device processes. So new facilities must be compatible with both biological and semiconductor best practices. In particular, this poses interesting challenges for contamination control and safety.

The design and engineering challenges for nanotechnology facilities include:

- Establishing and maintaining critical environments: temperature, humidity, and pressure
- Structural vibration isolation
- Airborne vibration isolation (acoustic noise)
- Isolation of mechanical equipment-generated vibration/acoustic noise
- Cost-effective power conditioning
- Grounding facilities for low electrical interference
- Electromagnetic interference (EMI)/Radio frequency interference (RFI) isolation
- Airborne particulate contamination

- Airborne organic and chemical contamination
- Environment, safety and health (ESH) considerations
- Flexibility strategies for nanotechnology facilities

The remainder of this document addresses the technology challenges unique to each of the areas mentioned above and provides best practices and examples of engineering approaches to address these challenges. It presents the ideas that were generated by the various breakout groups and supplemented by contributions from professional building design engineers. The objective is to provide insights on the design priorities and trade-offs for buildings to be utilized for nanotechnology, and other environmentally sensitive, research.

Chapter 2
Design Criteria

Abstract In the context of advanced technology laboratories, the "environment" is defined as the surrounding conditions under which sensitive laboratory equipment must be maintained for optimal performance. The cutting edge science of measurement and manipulation on an atomic scale requires extraordinary environmental stability. In large measure, many of the nanotechnology research accomplishments are due to facilities built to meet the stringent requirements compelled by nanometer-scale research programs. This chapter addresses the state of emerging guidelines and recommended practices, and discusses the array of technical and human criteria that emerge in the design of laboratories and cleanrooms supporting nanoscale science and technology laboratories. It sets the stage for the subsequent chapters.

Introduction[1]

Nanotechnology is no longer a new emerging field. In fact, nanotechnology has been present on scientists' radar for over three decades. Particularly in the last two decades, there have been amazing strides in nanotechnology research and product development. Many of these accomplishments are due in large part to the facilities built to meet the stringent requirements presented by many nanometer-scale research programs. Existing design and construction guidelines do not fully address the requirements of this new facility type. This chapter addresses the state of emerging guidelines and recommended practices, and discusses the array of technical and human criteria that emerge in the design of laboratories and cleanrooms

---

[1] Information in this chapter was originally published in "NanoTalk," a regular column authored by Ahmad Soueid for *Controlled Environments* magazine, http://www.cemag.us. Text and graphics from those columns are reprinted in this report with the permission from Vicon Publishing, Inc., *Controlled Environments* magazine. Please see "Additional Reading" following the References section at the end of this chapter for a complete listing of additional reading material.

© Springer International Publishing Switzerland 2015   5
A. Soueid et al. (eds.), *Buildings for Advanced Technology*, Science Policy Reports,
DOI 10.1007/978-3-319-24892-9_2

supporting nanoscale science and technology laboratories. Subsequent sections in this report focus on specific topics and/or present individual building case studies.

Look for the New IEST Guidelines

During the past few years, a number of nanotechnology facilities have been constructed and are now fully operational. During this time, the Institute of Environmental Sciences and Technology (IEST) has derived important practical information from these nanotech facilities. IEST's Working Group NANO-200 is dedicated to publishing the first set of guidelines that address nanotechnology research and production facilities. These guidelines address planning, design, construction, and operational considerations for facilities engaged in research or production at the nanometer scale.

Establishing Technical Requirements for a Building Project

Establishing technical requirements is the important first step to designing a building that will not become prematurely obsolete. Nanoscale research is a vast and complicated field that can engage a wide range of disciplines from both basic and applied sciences. Inherent in the nature of emerging areas of research, coupled with program and physical location constraints, are requirements that make each building project unique: there is no standard design solution. It would be nice to be able to build a typical laboratory and command that "thou shall conduct nanoscale research here." In reality, many types of nanoscale research require specialized laboratories designed to meet an array of special criteria. It is not unusual for conflicts to arise in the technical criteria as spaces are integrated. Some spaces may need an environment clear of airborne particulates; others require severely limited vibration and noise; while stringent temperature and humidity control must be maintained in others. A problem arises, for example, when the strict temperature and humidity control required in one area necessitates the use of large air handling equipment, which negatively affects the stringent vibration and noise limits required in other areas.

   The complex technical requirements for nanoscale research are aimed at creating an interior environment that is controlled, to a degree far exceeding other research facilities, for proper operation of the highly sensitive equipment required. See Fig. 2.1 for a comparison of the requirements imposed for the different types of laboratories. The technical considerations might involve controlling temperature and humidity levels, isolating vibration and acoustic noise, reducing the quantity and size of airborne particulates, and shielding from electrical, electromagnetic, and

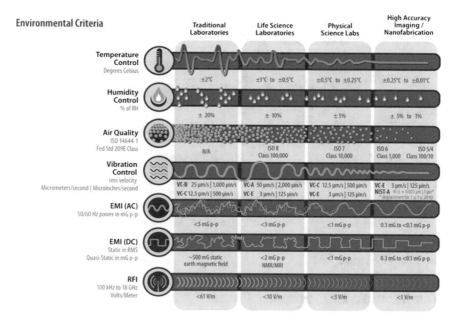

Environmental Criteria	Traditional Laboratories	Life Science Laboratories	Physical Science Labs	High Accuracy Imaging / Nanofabrication
Temperature Control Degrees Celsius	±2°C	±1°C to ±0.5°C	±0.5°C to ±0.25°C	±0.25°C to ±0.01°C
Humidity Control % of RH	± 20%	± 10%	± 5%	± 5% to 1%
Air Quality ISO 14644-1 Fed Std 209E Class	N/A	ISO 8 Class 100,000	ISO 7 Class 10,000	ISO 6 ISO 5/4 Class 1,000 Class 100/10
Vibration Control rms velocity Micrometers/second | Microinches/second	VC-B 25 µm/s | 1,000 µin/s VC-C 12.5 µm/s | 500 µin/s	VC-A 50 µm/s | 2,000 µin/s VC-E 3 µm/s | 125 µin/s	VC-C 12.5 µm/s | 500 µin/s VC-E 3 µm/s | 125 µin/s	VC-E 3 µm/s | 125 µin/s NIST-A VC-E + 0.025 µm | 1 µin* * displacement for 1 to f ≤ 20 Hz
EMI (AC) 50/60 Hz power in mG p-p	<5 mG p-p	<3 mG p-p	<1 mG p-p	0.3 mG to <0.1 mG p-p
EMI (DC) Static in RMS Quasi-Static in mG p-p	~500 mG static earth magnetic field	<2 mG p-p NMR/MRI	<1 mG p-p	0.3 mG to <0.1 mG p-p
RFI 100 kHz to 18 GHz Volts/Meter	<61 V/m	<10 V/m	<3 V/m	<1 V/m

Fig. 2.1 Environmental criteria for various types of laboratories (courtesy HDR Architecture, Inc.)

radio frequency interference.[2] Electron microscopes are most sensitive to many of the above "noise" sources. Temperature fluctuation and resulting changes in air properties along the length of a laser beam can cause beam distortion compromising accuracy in measurement.

The National Physical Laboratory (NPL) in Teddington, UK, houses various laboratories each requiring different environmental criteria be met. In this example, these technical requirements have guided the placement of individual components, heavily influencing the overall campus master plan, which includes a network of interconnected buildings designed as separate "modules" (see Fig. 2.2). Each building module is dedicated to a particular measurement science; each measurement science requires a high level of control over specific environmental criteria. This has resulted in a group of buildings that look similar on the outside but are specifically designed and engineered for particular functions on the inside.

---

[2] EMI(DC) in Fig. 2.1 refers to variation in otherwise static fields (such and the earth's magnetic field) due to time-varying changes in the environment (opening/shutting a nearby door that contains ferrous material, for example).

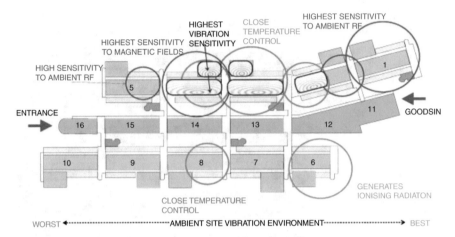

Fig. 2.2 National Physical Laboratory, Teddington, United Kingdom site criteria diagram (courtesy HDR Architecture, Inc.)

Temperature and Humidity Control

Achieving a specific temperature and keeping it constant can be more difficult than it seems. Work conducted in a lab can change the ambient air temperature enough to interfere with the experiment. However, by designing systems that circulate air inside the room with a relatively high number of air changes per hour, temperature can be closely controlled and maintained. In typical laboratories, temperature fluctuations are controlled to ±1 °C. When closer temperature control is needed, the tolerance can range from ±0.5 °C down to ±0.01 °C in highly critical spaces. Similarly, humidity control, depending on research requirements, may vary between ±5 % down and ±1 % of relative humidity (RH).

The National Institute of Standards and Technology (NIST) Advanced Measurement Laboratory (AML) provides an excellent example. Baseline average room-air temperature for all laboratories is controlled to within ±0.25 °C. There are 48 precision temperature control laboratories, located in two underground metrology wings, requiring temperature control to within ±0.1 °C or ±0.01 °C and humidity control to 1 % of RH. These laboratories are designed to house highly specialized experiments and equipment, such as the Molecular Measuring Machine (M3), an instrument designed to measure the dimensions and spacings of lines on silicon as small as 10 nm wide and 1 nm high (see Fig. 2.3). The Atomic Scale Characterization and Manipulation Laboratory has a system for atomic scale measurements of the electronic and magnetic properties of materials that uses a cryogenic (4 K), high magnetic field (10 T) scanning tunneling microscope which can be employed to build and measure nanostructures at the single-atom scale. Arrays of high-precision thermistors and humidity sensors are individually calibrated by NIST researchers to control the temperature, ensuring a tolerance of less than ±0.003 °C. Such extreme accuracy is critical to the function of the precision direct-digital-control systems

Fig. 2.3 NIST Advanced Measurement Laboratory I molecular measuring machine (courtesy of Steve Hall © Hedrich Blessing 2004)

serving the AML laboratories, to maintain the environments to meet the specified criteria.

Vibration Isolation

When measuring or capturing an image at the atomic scale, the slightest vibration can invalidate results; extremely small motions can distort images created by an electron microscope. First, the level of vibration that can be tolerated must be determined and the potential site tested at the outset to gauge ambient vibration levels, both during the day and at night. Once the threshold is determined and the site's vibration level is identified, then building design options can be considered. The more stringent the requirement, the more expensive the structure will be. A solution could be as simple as locating the building an adequate distance away from the street to minimize vibration caused by traffic. However, not all building sites allow for such deep setbacks.

The Sandia National Laboratories Center for Integrated Nanotechnologies (CINT) in Albuquerque, NM, is an example where the building was located outside the perimeter of the main campus in order to provide more distance between the building and external noise environment. This isolated the building from potential noise generated from surrounding buildings, and it provided greater distance between the facility and vehicular traffic. Furthermore, quiet and ultraquiet laboratories were required within the same facility. To accommodate the technical specifications, the Sandia facility was designed with three wings. The most vibration-sensitive laboratories were located the furthest from the wing where the research would generate the most noise.

The NIST Advanced Measurement Laboratory (AML) was located in a specific area of the NIST campus to resolve a vibration issue. The vibration criterion required baseline velocity amplitude of 3 μm/s and down to 0.5 μm/s or less in 27 low-vibration modules. The latter is 15–100 times better than in NIST's existing general-purpose laboratories. Two sites were available. Originally, one of the potential sites, the North site, was preferred because of its prominence and visibility at the entrance to the campus. Unfortunately, the vibrations at the preferred site were found to exceed tolerable levels (shown as curve "A" for NIST-A's North and South sites in Fig. 2.4). The greatest vibration at the South site (which was ultimately selected) was within the tolerable range, and furthermore, occurred at 4:00 a.m. when a train passed by on tracks over a mile away. The South site was chosen, although less prominent, because its characteristics are much better suited to the research planned to be conducted there.

In addition to choosing the quieter site, vibration was further reduced by building two of the facility's five wings underground, minimizing the effects of predominant surface vibrations. With the sensitive area about 13 m below ground level, floor vibrations were reduced by more than 50 % (see Fig. 2.5). For specialized laboratories, where limiting vibration is essential, it often makes sense to build underground.

But the "right" site may not be enough. For ultra-quiet spaces, rooms can be built on specially designed isolation slabs that are supported on a series of isolators, reducing the transmission of vibrations from the building to the slab where the experiment is mounted. In this case, the isolation slab is made of a concrete mass

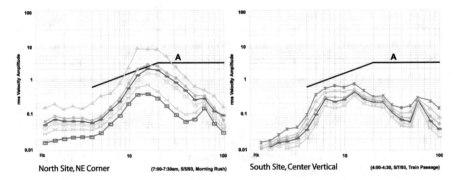

Fig. 2.4 The two graphs show six comparable sets of day-long vibration measurements for two potential building sites (courtesy of HDR Architecture, Inc.)

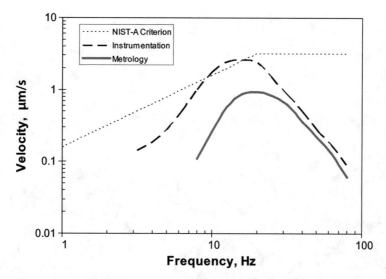

Fig. 2.5 NIST Advanced Measurement Laboratory I comparison of vibration measurements: two identical slab constructions located in instrumentation and metrology wings reflect the clear vibration isolation benefits in the slabs built 12 m below grade (courtesy of Colin Gordon and Associates)

weighing about 10–30 tons and it is set on air spring isolators. The isolation slab is suspended about a 6–7 mm above the base, ensuring that people working in the lab and walking across the floor will not cause the instrument to vibrate. It is important to accurately assess the level of vibration that the research can accommodate to avoid "over-designing" the lab, which is not cost-effective, or "under-designing" the lab, which can compromise results.

Airborne Contamination

At the nanometer scale, some spaces may require the control of airborne particulates (dust particles). Others may require the control of molecular contamination (biological matter). These two forms of airborne contamination are distinct and require different engineering and architectural solutions. These solutions may include segregation of spaces, in addition to air management systems that will provide positive or negative pressurization in the space where research or fabrication is conducted.

The Birck Nanotechnology Center at Purdue University is an example of a higher education facility built to include both nanofabrication in a traditional cleanroom environment and a pharmaceutical-grade cleanroom for molecular-based bioscience research (Fig. 2.6). While both spaces are within the same

Fig. 2.6 Purdue University, Birck Nanotechnology Center I cleanroom (courtesy of Steve Hall © Hedrich Blessing 2008)

building envelope, the engineering solutions and protocols are customized for each unique environment.

EMI, RFI, and Electrical Power Conditioning

Many of the instruments used for nanoscale research are highly sensitive to electrical and electromagnetic interference. Everything from a small fluctuation in the electrical power supply to electromagnetic interference (EMI) or radio frequency interference (RFI) can degrade research. At the outset of design, it is important to understand the specific requirements of the anticipated research in order to determine if EMI/RFI or power supply variation could be problematic.

Sometimes, the source of interference comes from forces outside the building. An outside interference issue arose during the early planning phases at the Center for Functional Nanomaterials at Brookhaven National Laboratory (BNL). Concerns arose regarding the radio frequency emissions produced by Doppler radar located about a mile away. The site was tested specifically for that issue. Fortunately, the RFI was low enough that it could be handled with local attenuation.

From an interior perspective, designers working to limit interference must be cognizant of how different types of labs are arranged. For example, labs that are particularly sensitive to electromagnetic interference should not be located near elevators or lab equipment that will produce strong electromagnetic fields. Certain systems necessary for maintaining the building must be taken into account. Transformers, rotating equipment, and pumps are examples of equipment required to keep basic building systems operational, but that can also wreak havoc with laboratories. These essential building systems should be located where their effects are as benign as possible, and if necessary, local shielding can help create an environment appropriate for nanoscale research. Generally, when designing the interior of a nanoscale research facility, components known to cause interference should be segregated from research areas as much as possible.

Planning for Facility Maintenance[3]

All facilities require regular maintenance. However, buildings that provide environments for advanced technology impose unique requirements for sustaining equipment. These facilities are increasingly required to operate on a 24 h, 7 days a week schedule. Additionally, relevant instrumentation and equipment have become automated allowing long-term experiments and measurements to be performed. As a result, even 1 day or so interruptions to accommodate maintenance requirements can cause problematic if not unacceptable disruptions for long-term experiments. These interruptions become particularly important for systems that provide high-level of temperature control. For example, a 1-day maintenance interruption in temperature control systems where a 1 mK tolerance must be sustained may require a week or more to reestablish acceptable temperature control levels. Many advanced coordinate measuring machines have measurement volumes of up to 1 m^3 with resultant machine masses of many tons—usually of cast iron. Achieving temperature stabilization of the environment and enclosed massive machines requires significant time. For this reason, maintaining back-up replacement components for critical components in temperature control systems is critical. Critical component backup for other environmental controls i.e., air-filtering systems for clean rooms should likewise be maintained. For similar reasons, planned operating disruptions to accommodate maintenance requirements should be closely coordinated with all users of the facilities to minimize disruptions to dependent experiments.

Safety

It is important to pay attention to environment, safety, and health considerations when designing an advanced research facility. Issues that must be addressed can be found in the presentation, "Safety in the Nano Research Laboratory Nanoscale

---

[3] Thanks to Dr. Theodore Doiron for recommending the inclusion of this topic.

Science and Engineering Laboratory Buildings."[4] Other pertinent documents are the March 2009 DHHS (NIOSH) publication No. 2011-206, "Approaches to Safe Nanotechnology: Document Provides Guidance to Protect Nanotechnology Workers,"[5] and the May 2012 DHHS (NIOSH) publication No. 2012-147, "General Safe Practices for Working with Engineered Nanomaterials in Research Laboratories." As an example of policy/guidance to facility users, see the May 2008 Department of Energy Nanoscale Science Research Center's, "Approach to Nanomaterial ES&H."[6]

Sustainability

Sustainable design is an asset to advanced research facilities. First, sustainable design is holistic in its approach, that is, it promotes true consideration and accounting of overall building performance and impacts. Its nontraditional design outcomes, such as occupant health and productivity, or gallons of water conserved, are objectives that are calculated from the beginning of the design process and that are tracked during the life of the building. Just as metrics are critical to any advanced research, so too are they critical for sustainable design and development. Second, in almost every instance these sustainability goals enhance the research environment. Whether it is better air quality, carefully day-lighted office space, reduced energy demand, or even operations and maintenance savings to the owner, sustainability supports advanced research environments.

Designing sustainable features into buildings for nanotechnology R&D can be a daunting proposition: there are already so many technical design criteria that must be addressed. However, sometimes, sustainable features work in tandem with the technical criteria. For example, at Purdue University's Birck Nanotechnology Center, the sustainable landscape design includes indigenous grasses that do not require mowing. This not only reduces maintenance and energy costs, it also eliminates the vibration disturbance caused by mowers. These types of relatively simple design decisions can have a significant effect on the sustainable rating of advanced research facilities.

The most widely recognized systems for "green" assessment of building design are the UK's Building Research Establishment Environmental Assessment Method (BREEAM),[7] established in 1990, and the newer U.S. system, the Leadership in

---

[4] Delivered at the December 2012 NSF NSE Grantees Conference by Mark Jamison of HDR Architecture, Inc. The presentation is available at the conference website http://www.nseresearch. org/2012/program.htm#Day1

[5] http://www.cdc.gov/niosh/docs/2011-206/

[6] http://orise.orau.gov/ihos/Nanotechnology/nanotech_DOE_Nanoscale_SC.html and http://nano. anl.gov/events/workshops/enm/index.html

[7] http://www.breeam.org/. BREEAM is the world's leading design and assessment method for sustainable buildings.

Energy and Environmental Design (LEED),[8] launched in 1998. Canada also has an established sustainability program called Golden Globes,[9] and other similar programs are available across the world. The process of sustainable development, from design inception through the end of a building's useful life, is a sound way to reduce facility costs, improve the working environment, and demonstrate good stewardship to the public. Sustainability and advanced research facility design is a win–win relationship.

Researchers Are People, Too!

Although the complex technical requirements discussed above are paramount, designers of facilities for nanotechnology research must also evaluate and make provisions for human criteria. New research facilities are often used as a recruitment tool to attract top scientists to an institution. The present and future scientists who conduct research within the facility, as well as the interface between the facility and the community at large, are factors that cannot be overlooked.

It is crucial to remember that a technically "perfect" nanoscale laboratory will not function well if it is located in a facility where researchers do not want to work. Designing spaces for nanotechnology research is not only about how atoms, materials, and machines interact, it is also about how people interact. New ideas, new techniques, and new products result from people working together with one another. Buildings can facilitate such interactions through the design of welcoming and comfortable spaces.

The National Physical Laboratory in the UK is an example of how technical criteria can be isolated without isolating the scientists. In this facility, various laboratories are designed as separate modules, yet the laboratories are connected through a series of bridges and links to shared spaces that allow for communication, collaboration, and collective activities.

Another example of providing for interaction among scientists is the Sandia National Laboratories and Los Alamos National Laboratory Center for Integrated Nanotechnologies. The facility was intentionally built outside the perimeter of the main campus, allowing researchers from different countries to collaborate with greater ease. The new building is secure, yet the protocols are less stringent than would be required for work within the perimeter of the main Sandia campus.

The Brookhaven National Laboratory Center for Functional Nanomaterials contains highly sensitive laboratories and cleanrooms while providing ample opportunity for collaboration among scientists in strategically placed interaction areas. These interaction areas are also found in other buildings like Purdue University's Birck Nanotechnology Center (see Fig. 2.7). Well-located and well-

---

[8] http://www.leed.net/, "Promoting LEED Certification and Green Building Technologies."

[9] http://www.theglobeawards.ca/home: Globe Foundation awards for environmental excellence.

Fig. 2.7 Purdue University, Birck Nanotechnology Center I interaction area (courtesy of Steve Hall © Hedrich Blessing 2008)

used interaction areas are essential to fostering the impromptu communication inherent in interdisciplinary research.

Public Interface

A building's iconic image, helpful in attracting researchers, is also a tool for communicating the facility's function to the public at large. The Joint School of Nanoscience and Nanoengineering in Greensboro, North Carolina, illustrates how extra care was devoted to using the building as a tool for communication (Fig. 2.8). Transparency in the design allows people to see inside; a clear indication of investment in the local economy and commitment to the local community. In addition, the distribution of public space within the building provides maximum, yet safe, exposure to foster public awareness.

Fig. 2.8 Joint School of Nanoscience and Nanoengineering, Greensboro, NC (courtesy of HDR Architecture, Inc.)

Fig. 2.9 Nanoscale Material Science and Engineering Building at the University of Arkansas (courtesy of HDR Architecture, Inc.)

Designing for the Unknown (the Future)

The Nanoscale Material Science and Engineering Building located at the University of Arkansas in Fayetteville is another example of a facility designed with the future in mind (see Fig. 2.9). The university made its initial capital investment in a building designed to allow its research program to grow. In addition to flexibly designed laboratories and cleanrooms, the university created open shelled spaces that can be completed and adapted as funding becomes available and research parameters ascertained.

Conclusion

Both the technical criteria and human criteria that must be addressed in the design of facilities for nanotechnology R&D can present astounding complexities. Even the best planning effort cannot predict perfectly how spaces will be used, or anticipate technological advancements that may dramatically affect space requirements. Over time, spaces will be assigned to different researchers and programs. Flexibility and adaptability features are critical to any design. Just as interdisciplinarity is evolving and shaping the future of nanoscale research, architecture and building technologies in this context are also evolving. Traditional methods and models must yield to new partnerships among architects, engineers, nanoscale researchers, technical experts, and institutions' representatives. Combining and integrating the perspectives and expertise of all parties involved will culminates in a sound basis of design for the future.

Acknowledgement Thanks to Ahmad Soueid, HDR, who integrated materials from various sources into the initial draft of this chapter.

Bibliography

H. Amick (ed.), Buildings for nanoscale research and beyond, in *SPIE Proceedings*, vol. 5933, 2005
H. Salem, Nanotechnology research center: designing facilities that support nanotechnology research at university, PhD thesis in Architecture Engineering, Alexandria University, Egypt, July 2011
A. Soueid, H. Amick, T. Zsirai, Addressing the environmental challenges in Nano fabrication and measurement laboratories. Presented at buildings for nanoscale research and beyond, a SPIE conference, San Diego, CA, 31 July–1 Aug 2005
A. Soueid, National Institute of Standards and Technology Advanced Measurement Laboratory, published by HDR, 2005
A. Soueid, Sandia National Laboratories, Center for Integrated Nanotechnologies (CINT) Core Facility, published by HDR, 2008
A. Soueid, Chapter 6—Designing for the future: nanoscale research facilities, in *Presenting Futures*, ed. by E. Fisher, C. Selin, J.M. Wetmore. The Yearbook of Nanotechnology in Society, vol. 1 (Springer, Dordrecht, 2008). http://www.springerlink.com/content/k725288041428874/
A. Soueid, Nanotechnology research requires a different breed of laboratories and cleanrooms, *Controlled Environments Magazine*, December 2009

A. Soueid, The new breed of nanotech facilities, *Controlled Environments Magazine* (Vicon, December 2009)

A. Soueid, nanoTALK: NIST advanced measurement laboratory complex, *Controlled Environments Magazine* (Vicon, January 2010)

A. Soueid, nanoTALK: show me the (NANO) money, *Controlled Environments Magazine* (Vicon, May 2010)

A. Soueid, nanoTALK: now we're talking! (a conversation with Dr. Clayton Teague), *Controlled Environments Magazine* (Vicon, September 2010)

A. Soueid, nanoTALK: nanotechnology is 10 years old—what's next?, *Controlled Environments Magazine* (Vicon, April 2011)

A. Soueid, nanoTALK: i'm pickin' up (NANO) vibrations... (the building) is giving me excitations..., *Controlled Environments Magazine* (Vicon, September 2011)

A. Soueid, nanoTALK: nano is big in North Carolina, *Controlled Environments Magazine* (Vicon, January 2012)

Chapter 3
Temperature and Humidity Control

Abstract The instruments to manipulate atoms or observe their motions require stricter environmental specifications than prior generations of equipment. The primary focus of this chapter is temperature and humidity control, two of the most critical variables that affect work at the nanometer scale. In basic science laboratories, environmental control to an accuracy of ±0.5 to $0.25\ °C$ and ±5 to $10\ \%$ relative humidity (RH) is adequate. For atomic-scale measurements, a much more highly controlled environment is required. Because test equipment is extremely sensitive to temperature and humidity variations, ±0.1 to $0.01\ °C$ and as close to $\pm1\ \%$ RH as possible is needed. The preferred standard air condition for nanoscale research is $20\ °C$ at 40–$45\ \%$ RH. This chapter addresses the various design tradeoffs and control systems needed to accomplish these standards.

Introduction

Nanotechnology research brings together participants from a wide variety of scientific fields including basic and applied sciences, engineering, semiconductor design and fabrication, and advanced computational analysis. Laboratories used for nanotechnology research require accommodating these disparate disciplines and maximizing opportunities for integration, while avoiding the adverse effects that various R&D processes can have on one another. In addition, the cutting edge science of measurement and manipulation on an atomic scale requires extraordinary environmental stability. In the context of advanced technology laboratories, the "environment" is defined as the surrounding conditions under which sensitive laboratory equipment must be maintained for optimal performance. The primary focus of this chapter is temperature and humidity control, two of the most critical variables that affect work at the nanometer scale.

© Springer International Publishing Switzerland 2015

A. Soueid et al. (eds.), *Buildings for Advanced Technology*, Science Policy Reports,
DOI 10.1007/978-3-319-24892-9_3

Environment Control for Instrumentation

There are a number of reasons why it is necessary to maintain strict control over conditions surrounding sensitive research equipment. For metrology instrumentation in particular, the main objective is to rule out as many variables as possible as measurements are being taken. Increased resolution in microscopy and nanoprobe tools is driven by the need to manipulate atoms or observe reactions in real-time, for example, in development of new nanoscale catalyst materials. This need is also resulting in stricter environmental specifications for these types of instruments. In response to these requirements, manufacturers of equipment used in nanotechnology research have greatly tightened their environmental specifications. For example, one manufacturer of transmission electron microscopes (TEMs) requires the following ambient conditions (Table 3.1).

These requirements are made all the more difficult to achieve because the equipment itself is a heat source, as are the operators if the equipment is not set up for remote operation.

Laboratories dedicated to nanotechnology research differ from other laboratory types, such as wet chemistry or biology laboratories, because of the increased focus on strict control of the instrument environment. In basic science laboratories, environmental control to an accuracy of ± 0.5 to 0.25 °C and ± 5 to 10 % relative humidity (RH) is adequate. For atomic-scale measurements, a much more highly controlled environment is required. Because test equipment is extremely sensitive to temperature and humidity variations, ± 0.1 to 0.01 °C and as close to ± 1 % RH as possible is needed. The preferred standard air condition for nanoscale research is 20 °C at 40–45 % RH.

Design Challenges

Temperature

A designer of critical environments faces an initial challenge to determine, with the scientists who will potentially use the laboratory, the definition of stability. Stability of variables such as temperature and humidity can mean stability over time (drift or fluctuation), over space (uniformity), or both. The range of solutions varies significantly depending on the requirement.

Table 3.1 Illustrative TEM requirements

Temperature	20 °C ± 0.25 °C to ± 0.01 °C
Drift	0.05 °C/h
Fluctuation	0.005 °C/min
Air velocity	Less than 5 m/min (40 fpm)

m/min meters per minute, *fpm* feet per minute

For example, stability over time can be addressed by attention to control systems, including the sequence of operation, selection, and location of sensors, and by layout of the laboratory equipment in the space. Uniformity can be achieved by increased air change rates; by proper distribution, supply, and return of air; and through equipment modification, and location and installation details (e.g., locating equipment out of the path of supply air and providing perforated tables). Increased airflow can also have a positive impact on temperature stability by reducing the air temperature difference across the room, providing the opportunity for closer control.

It is important to understand how equipment will be used (i.e., whether the equipment will be controlled from a remote location or by operators within the space). The impact of the equipment itself also requires evaluation, since it contributes to the heat load. It is also necessary to determine how the operator or equipment loads vary over time, in order to gain an understanding of the response time that the control system will be required to achieve.

Finally, consideration must be given to diversities in equipment usage and load and to potential future uses of the proposed critical environment laboratory, to ensure that the final design can accommodate future needs without significant retrofit.

Air Filtration

In order to meet cleanliness criteria, air-handling units have several levels of filtration. Pre-filters are used to protect components of the units, especially the cooling coils. These are preceded by panel filters to extend the pre-filters' lives. At discharge, upstream humidifier units serving laboratories have 95 % cartridge or 99.997 % HEPA filters. Additionally, many air units, particularly those in cleanrooms, require carbon filters to assure removal of gaseous contaminants. Air handlers serving noncritical spaces do not normally require this high level of cleanliness, although upgraded filtration is sometimes requested by those who are concerned about maintenance and the small particles resulting from wear on the fan belt.

Humidity

One special feature of the nanotechnology laboratory is its requirement for relatively low humidity, 40–45 % RH at 20 °C. This RH is low enough to avoid corrosion and high enough to minimize static charge accumulation. Since some of the laboratory equipment is water cooled, low humidity is essential to prevent condensation within the instruments. Furthermore, most test equipment is calibrated at the above conditions as they have become an international standard for this type of work. In the traditional HVAC (heating, ventilating, and air conditioning) system, air is cooled to 10 °C, reheated by the fan to 12–13 °C, and distributed to the space. With a space temperature of 20 °C, this would result in humidity close to 60 % RH, too high to avoid condensation. To overcome this, a secondary cooling coil using −1 °C glycol solution supplies 5–6 °C air, providing additional dehumidification. In order to avoid any condensation in the distribution system, this air is reheated to 12–13 °C before leaving the air handling unit.

Air Management

When uniformity of temperature over a specific volume is a requirement, it is necessary to provide sufficient air to the space to remove the heat load and minimize the temperature rise. The airflow required (as measured by air changes per hour) is a function of both the sensible heat gain and the maximum allowable temperature gain (variance from setpoint). For example, for an average heat gain of 1000 W (25 W/m^2 for a module 4 m by 10 m with a 3 m ceiling height), approximately 280 air changes per hour would be required to maintain ±0.1 °C temperature uniformity throughout the space. Achieving a temperature uniformity of ±0.01 °C requires either a significant number of air changes (well in excess of 280 per hour) or a significant limitation on heat gain into the controlled space. By comparison, the number of air changes per hour in a typical Class 100 cleanroom ranges from 250 to 600 per hour.

There are a number of other variables that must be considered when estimating heat gain for the purposes of calculating required air flow, including equipment diversity and heat rejection to process cooling water or a dedicated exhaust stream.

Equipment manufacturers also frequently place restrictions on air velocity to minimize air currents next to critical components. Airflow around the column of an electron microscope can cause slight vibrations that affect the image. For the example described above, an air change rate of 280 per hour in a 4-m by 10-m laboratory module results in an average air supply velocity of approximately 15 m/min, assuming distribution over the entire ceiling area. Since this is not usually the case, the volume delivered at each diffuser increases significantly depending on the number of diffusers allocated. To meet target air velocity limits, the designer can specify diffusers specifically designed for high-volume, low-velocity applications and coordinate diffuser placement with the equipment location.

Control System

Regardless of the degree of stability required for the critical environment, there are several issues that must be addressed in design of the HVAC control system. These include deciding which variables will be controlled, selecting and locating the sensors, and developing a sequence of operation that will accomplish the desired effect. Determining which variables will be controlled involves the broader decision about how the laboratory air supply will be zoned, since it is likely that different spaces will have different setpoints and control range. It may be sufficient to control humidity in some areas at the makeup air-handling unit, while other areas may require local zone humidification control in order to achieve a tighter range of control. Once the air supply zoning strategy is established, the designer can identify the control hierarchy for each of the controlled variables, such as temperature and humidity.

Selection and location of sensing devices is also critical. Thermistors are frequently specified for high accuracy measurements since they have a higher sensitivity to temperature changes than resistance temperature detectors (RTDs). However, thermistors have a nonlinear response and are typically used where the measured temperature range is relatively narrow. Thermistors are well suited for measurements in spaces where rapid response to changes is needed. Sensors should be located in areas within critical laboratories that are not influenced by highly variable load sources (e.g., away from non-steady-state heat generating equipment). Sensors are typically located in the proximity of the critical equipment; however, if the heat load introduced into the space varies relative to the sensor location, the thermal lag between source and measuring point needs to be identified and accommodated by the control system.

Vibration Controls

The needs for cleanliness, clean power, an EMI/RFI-free environment, and accurate temperature and humidity control differ with the type of work being conducted, but most nanotechnology research activities are sensitive to vibration. To meet the stringent criteria this imposes, all fans, pumps, and other rotating equipment are mounted on carefully selected spring isolators, as are all ducts and active pipes. The fluid velocity in ducts and pipes is limited to a safe level. Industrial-grade air handling units with 100 mm walls and double vibration isolation are used. Large ducts have heavier gauge walls than those required by the Sheet Metal and Air Conditioning Contractors' National Association. Air handlers, exhaust fans, and air terminal units have sound attenuation. In general laboratories, diffuser and register sizes are selected to limit their noise levels to NC 30[1] or lower.

---

[1] An acoustic noise criteria standard; see Chap. 5 for more details.

Energy Efficiency

All laboratories are large energy users, and nanotechnology labs are especially demanding in this respect. In general laboratories, a large volume of air must be circulated to offset the heat generated by the laboratory equipment, and a large proportion of outside air must be used to make up fume hood exhaust. In cleanrooms and high accuracy temperature-controlled laboratories, a large volume of air is circulated by multiple air-handling units. The low relative humidity requirement demands additional cooling and reheat energy. In order to minimize energy consumption, various conservation techniques are used.

Coil loop heat recovery is installed in exhaust systems and air handling units serving laboratories. This utilizes heat from the warm exhaust to preheat cold outside air in the winter and cool the incoming warm air in the summer. While the system recovers sensible energy only, it allows for physical separation of the exhaust ductwork from the outside air to provide greater flexibility in the installation and assures that there can be no cross-contamination between exhaust and supply. In order to enhance the system's effectiveness and efficiency, the coils have air bypass capabilities to minimize fan use during non-recovery periods.

In air systems serving labs that require low relative humidity, waste heat is recovered from the low-temperature chiller and is used to reheat the air before it is distributed to the labs. To minimize fan and pump energy, variable volume air and hydronic systems utilize adjustable frequency drives where feasible. This allows precise balancing, operating flexibility, and most importantly, minimized energy cost.

Design of a laboratory intended to maintain critical environmental conditions must also take into account design of architectural and electrical systems as well as HVAC and controls. In addition, there are potential impacts on other aspects of a facility that can arise from solutions to temperature and humidity control.

One of the most important design considerations is sufficient space for air distribution. In order to meet the stringent acoustic criteria that are common for many critical environments (NC-25 or NC-30) and to minimize vibration transmission from ductwork to sensitive equipment, it is necessary to design ductwork with airflow velocities less than 250 m/min, or install sound attenuators, or both. This greatly increases the size of ductwork that must be installed to support the laboratory. Sufficient space around and above the laboratory is required for routing ductwork and for accessories, and these need to be coordinated with other services. Similarly, the large volumes of air that are distributed to laboratories with stringent uniformity requirements must be removed and returned, typically at the floor level. This minimizes temperature instability at elevations in the center of the room, where equipment is located. Sufficient floor area and unobstructed wall length need to be allocated for return air chases and grilles. Selection and placement of lighting fixtures also need to be coordinated with the equipment layout and sensor location, to minimize negative effects on HVAC control systems.

Integration of the Manifold Requirements

Individually, these criteria are not difficult to achieve. Cleanliness levels of better than Class 10 can be achieved. Vibration control technology for mechanical and electrical systems is well established. Clean and reliable electrical power and an EMI/RFI-free environment can be designed. Temperature can be controlled to ± 0.01 °C stability. The engineering challenge of the nanotechnology laboratory is to achieve all of these simultaneously. For example, we need to create an EMI/RFI free environment while controlling the fan speed with adjustable frequency drives. We need to control the temperature to 0.01 °C stability while minimizing vibration generated by air movement. At the same time, the design must be environmentally sensitive, minimizing energy consumption. This can be particularly difficult in a laboratory that requires 20 air changes per hour for cooling and 100 % outside air for safety, combined with the large electrical load of the circulating fans used to maintain the required cleanliness.

The most challenging task however, is to fit the systems and equipment into the available mechanical and electrical spaces. These spaces often appear limited when considering the many air handlers, exhaust fans, pumps, ducts, pipes, heat recovery systems, and electrical equipment that are needed for the safe, reliable, and efficient operation of these sophisticated buildings. The number of engineering systems can, in extreme cases, be five times more than would be required for a typical bio-research laboratory.

Complying with these frequently conflicting criteria requires close coordination within the design team and particularly with the user to identify the conflicts and limitations of the various systems. Priorities in the design requirements and the level of flexibility must be established.

General Laboratories

General laboratories usually have many fume hoods. Air handling units serving these spaces require a large amount of makeup air. Because of the potential hazard of contamination, recirculation is not an option, so the units must supply 100 % outside air. Even though some labs do not have fume hoods and air could be recirculated, 100 % outside air is used for future flexibility. Supply airflow must equal the largest of the exhaust makeup and lab cooling requirements. Since the complete list of initial and future equipment is typically unknown at the time of design, an estimated 100 W/m^2 equipment heat gain is used, and sufficient airflow is provided to accommodate this. Depending on the lighting, people, and external heat gain, the required supply air change rate for a general laboratory may vary between 15 and 20 air changes per hour. The required negative differential pressurization is maintained by exhausting more air than is supplied. See Fig. 3.1 for a typical lab air duct layout.

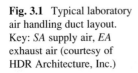

Fig. 3.1 Typical laboratory
air handling duct layout.
Key: *SA* supply air, *EA*
exhaust air (courtesy of
HDR Architecture, Inc.)

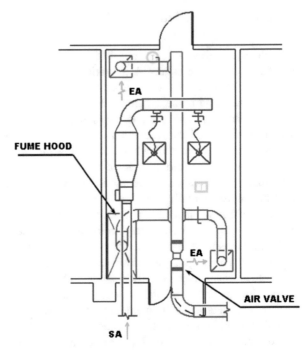

Each laboratory is served by a constant volume air terminal unit. Its function is to maintain constant flow to the room regardless of the pressure variation upstream in the ducts, and in this function it acts as an automatic balancing device. It also has a reheat coil that is used to control the room temperature by heating the supply air. The housing of the air terminal unit acts as a sound attenuator. Normally, a room sensor will maintain its setpoint by modulating the reheat coil hot water valve, varying the discharge temperature.

Where stability better than the typical ±1 °C is required, a more complex duct layout and control system is needed. In order to increase control stability, a temperature sensor is located in the duct downstream of the air terminal unit and controls the reheat coil to maintain a set discharge temperature. The setpoint of this control loop is automatically reset as necessary to maintain the desired temperature at the space sensor. This concept is effective in controlling to ±0.25 °C stability, but the temperature will be maintained only at the sensor location and not through-out the laboratory unless there is virtually no heat load or airflow is increased well above 15–20 air changes per hour. If the work requires close temperature control in only a limited area, a simple solution is to use a portable space sensor which may be moved to any work area for localized control.

An example of a laboratory designed to maintain ±0.25 °C (approximately ±0.5 °F) throughout the space is provided in Fig. 3.2. In this example, assuming a laboratory size of 4 m by 10 m by 3 m and 20 air changes per hour, a total of 16 high-entrainment-type supply air diffusers would be required to distribute the air

Fig. 3.2 Laboratory layout
for ±0.25 °C temperature
control. *RA* return air, *SA*
supply air, *HAC* high-
accuracy controller,
T temperature sensor
(courtesy of HDR
Architecture, Inc.)

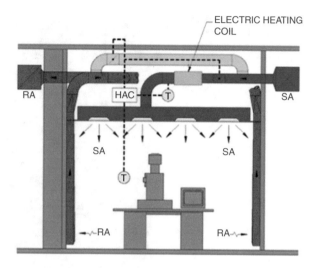

while keeping the supply velocity at 15 m/min or less. Low supply velocities are desirable to reduce airflow turbulence and the associated temperature change impacts on the equipment, and to reduce the acoustic impact on the lab. Return air can be ducted, as shown in Fig. 3.2, or through an adjacent chase with return air grilles in the sidewalls. Return air grilles are typically selected to limit velocity across the grille to 90 m/min across the core area of the grille (this is similar, but not the same as, the free area) to minimize pressure drop and noise. Return air ductwork is similarly sized. When equipment adds significant localized heat sources, it may be necessary to consider spot cooling to minimize need for atmospheric heat dissipation.

The layout in Fig. 3.2 illustrates an electric heating coil as the terminal reheat device. Electric heating coils offer considerably faster control response compared to hot water coils, although at a tradeoff in overall energy efficiency. A bypass is shown around this coil, to reduce air flow to the room when less cooling is needed. This example also shows installation of temperature sensors near the critical equipment and in the supply air duct; the latter is used to provide the primary control signal to the heating coil and coil bypass damper.

High Accuracy Laboratories

Most research can be done in laboratories with traditional environmental control. However, nanoscale measurements and research often require temperature control stabilities of ±0.1 °C or a control stability as high as ±0.01 °C. A design solution for a high accuracy laboratory module (control to ±0.01 °C throughout the space) differs from the solutions shown in Figs. 3.1 and 3.2. The primary difference is in the architectural design of the laboratory itself. As illustrated in Fig. 3.3, the

Fig. 3.3 Example of layout
for ±0.01 °C temperature
control design (courtesy of
HDR Architecture, Inc.)

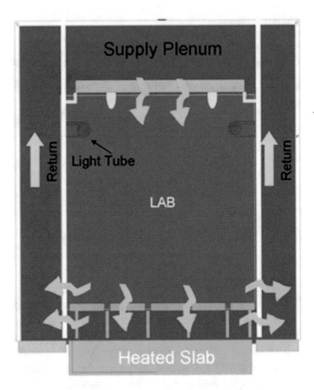

laboratory in this example is designed such that the return air path is through a
raised access floor and back through return air walls surrounding the room, in effect
isolating the lab from the surrounding spaces. This concept meets several objectives: a high rate of air supply can be efficiently distributed and returned, the room
is thermally isolated from its surroundings, and return air velocities are reduced to
improve acoustic and vibration performance.

Controlling the temperature in an entire laboratory space to ±0.01 °C stability
requires special engineering solutions. Due to the low specific heat of air, it is
difficult to remove heat generated within the space without significant temperature
increase, and such accuracy in a large space with more than minimal heat gain is not
easily achieved. Table 3.2 indicates the heat that can be introduced into a space at
various air change rates while staying within the required narrow band. In a typical
office with six air changes per hour, for example, the heat gain of an ordinary
flashlight will increase the room temperature by 0.02 °C, the span of the ±0.01 °C.
In contrast, if this narrow temperature range is to be maintained in the same space,
the heat generated by a human body requires 280 air changes per hour.

One conceptually simple solution is to supply the room with a large volume of
air, precisely controlled at discharge to ±0.01 °C stability. This discharge temperature then can be reset by one or more temperature sensors located strategically in
the space. As the air moves through the space, its temperature will increase as it
absorbs heat from the room. However, if the location and size of the heat source is

Table 3.2 Maximum
permissible heat gain

Air flow (l/s)	Air changes	Heat gain (W)
5000	280	123
4000	224	98
3000	168	74
2000	112	49
1000	56	25
500	28	12
100	6	2.5

controlled and the air flow rate is high enough, the temperature will stay within tolerance.

The air can be supplied into the lab by an above-ceiling plenum having perforated ceiling panels or HEPA filters. The advantages of the HEPA filters are that they provide excellent air distribution, ISO 5 (Class 100) cleanliness, and a soft ceiling for sound absorption; however, they are expensive. In a less costly solution, perforated ceiling tiles can be used in a configuration similar to a HEPA ceiling. This offers the opportunity to change the air distribution pattern and consequently the temperature gradient in the room by varying the number and size of the perforations in the panels. The disadvantage is the lack of sound attenuation. Low wall return air grilles continue the downward movement of air, drawing it away from the work area before it can remix.

In order to minimize the radiant heat transfer, all surfaces of the lab must be maintained close to the desired room temperature. Return air plenums within the double walls of the labs prevent inner wall temperature from being affected by adjacent labs (see Fig. 3.3).

Most of these labs are located on slab on grade. Depending on the climate, the slab may need to be heated. The heating can be done with a hydronic system, or with electric cables. The advantage of the hydronic option is that it can be more easily controlled and can be used for cooling if needed. The electric heating is less expensive but only an open loop control can be used effectively.

The room is equipped with work lights that produce the necessary lighting levels for setting up the experiments and measurements. Since these lights produce much more heat than can be tolerated during the actual testing, low heat emission light fixtures are used. These consist of a light tube in the room and a lamp located outside (see Fig. 3.4). Between the lamp and the light tube, an infrared filter minimizes radiated heat.

System Components

The high accuracy system includes a circulating air handling unit that typically provides as much as 300 air changes per hour. It has a prefilter, cooling coil,

Fig. 3.4 Low-intensity
lighting illustration
(courtesy of HDR
Architecture, Inc.)

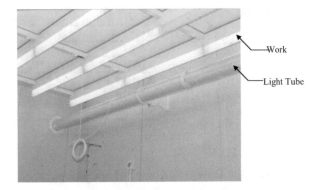

Fig. 3.5 Illustration of a
high accuracy air handling
unit. *RA* return air, *SA*
supply air, *VFD* variable
frequency drive (courtesy of
HDR Architecture, Inc.)

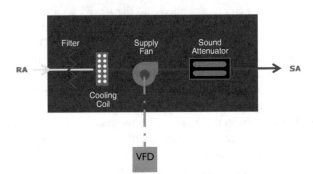

variable frequency drive (VFD)-controlled supply fan, and sound attenuator (see
Fig. 3.5).

The air distribution system (see Fig. 3.6) consists of four duct branches with
electric reheat coils in each. The four branches allow supplying the air to the above
ceiling distribution plenum evenly with minimum velocity and turbulence. The
bypass duct allows inactivation and complete isolation of the individual duct
branches when lower airflows are desired.

Another main component is the makeup air (MA) unit. Its purpose is to provide
humidity control and pressurization, and allow smoke evacuation in case of fire.
The recommended makeup airflow is six air changes per hour. This value is based
on a 10-min air change for smoke evacuation and is adequate to provide reliable
humidity control. For individual components, see Fig. 3.7.

Since these rooms are tightly built, they have their own exhaust fan sized to
remove the make-up air, less than that required for pressurization. For individual
pieces of heat-producing equipment, localized exhaust outlets or snorkels are
installed.

The heart of the control system is a low watt density electric reheat coil located
in each discharge duct. It is sized to heat the air to a maximum change of 1 °C at full
flow. A duct mounted high accuracy temperature sensor, using an electronic
modulating heat controller, maintains the temperature downstream to ±0.01 °C

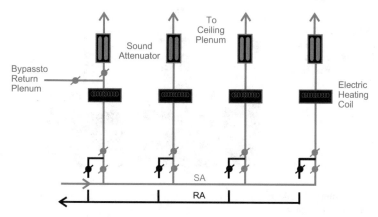

Fig. 3.6 Illustration of a four ducted air handling system. *SA* supply air, *RA* return air (courtesy of HDR Architecture, Inc.)

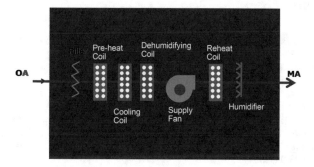

Fig. 3.7 Schematic of a make-up air handling unit. *OA* outside air, *MA* makeup air (courtesy of HDR Architecture, Inc.)

stability. The discharge temperature is reset by the room sensors through a high accuracy controller (HAC) to maintain room setpoint (see Fig. 3.8).

The control components include high accuracy temperature sensors and precision controllers. From the different types of sensors available, the best choice is the thermistor. Its advantages include high sensitivity, high resistance, fast response time, and low cost. The sensors need to be custom-calibrated to 0.001 °C accuracy. The high accuracy controllers are microprocessor-based digital, 14–16 bit input resolution, proportional integral derivative controllers with adaptive tuning capabilities. In order to give the scientists full and complete control over their work environments, a local interface panel is recommended. This allows them to set the room temperature, humidity, and airflow, and to dynamically observe the actual conditions. The interface panel should be mounted outside the lab and protected by a keypad for security.

Fig. 3.8 High accuracy
control schematic. *HAC*
high accuracy controller,
SCR silicon controlled
rectifier (courtesy of HDR
Architecture, Inc.)

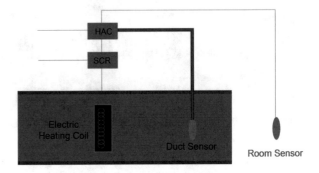

Acknowledgement Thanks to Ted Zsirai, who integrated material from various sources into the initial draft of this chapter.

Bibliography

V. Bradshaw, *The Building Environment: Active and Passive Control Systems* (Wiley, Hoboken, NJ, 2006). ISBN 978 0 471 68965 2

C. DiLouie, *Advanced Lighting Controls: Energy Savings, Productivity, Technology and Applications* (Fairmont, Lilburn, GA, 2005). ISBN 0 88173 510 8

R.W. Haines, D.C. Hittle, *Control Systems for Heating, Ventilating, and Air Conditioning* (Springer, New York, 2006). ISBN 0 387 30521 1

D. Heerwagen, D. Heerwagen, *Passive and Active Environmental Controls: Informing the Schematic Designing of Buildings* (McGraw Hill, New York, 2003). ISBN 0 07 250173 1

R. McDowall, *Fundamentals of HVAC Control Systems*, IP edn. (Elsevier, Boston, MA, 2009). ISBN 978 0 08 055233 0

Chapter 4
Vibration Isolation

Abstract Vibration control involves designing architectural, structural, mechanical, and electrical systems, both independently and in combination, to not generate or propagate vibration that is detrimental to research activities. The environment itself must be considered as an experimental variable and constrained to known and closely controlled values. For "routine measurements" the requirement is root mean square (RMS) amplitude displacement of 0.025 micrometer (μm) at frequencies between 1 and 20 Hz and RMS velocity amplitude of 3 μm/s at frequencies between 20 and 100 Hz. The metrology requirements are RMS velocity amplitude of 3 μm/s at frequencies below 4 Hz; RMS velocity amplitude of 0.75 μm/s at frequencies between 4 and 100 Hz. This chapter addresses the structural design of the building, the layout of the process and mechanical equipment, and the equipment layout in the laboratory or production spaces that are critical to achieving an acceptable level of vibration within specific areas.

Introduction

Vibration control involves designing architectural, structural, mechanical, and electrical systems, both independently and in combination, to not generate or propagate vibration that is detrimental to research activities.

In many areas of research, the environment itself must be considered as an experimental variable and constrained to known and closely controlled values. Advanced technology equipment is sensitive to vibration to varying degrees. For example, the most demanding metrology equipment requires that environmental vibration amplitudes be lower by a factor of 10 than those of typical microelectronics (photolithography) production floors, or lower by a factor of 100 than those of relatively conventional analytical laboratories. The vibration environments required by advanced research equipment generally are lower by a factor on the order of 1000 than vibration that can barely be perceived by humans.

© Springer International Publishing Switzerland 2015   35
A. Soueid et al. (eds.), *Buildings for Advanced Technology*, Science Policy Reports,
DOI 10.1007/978-3-319-24892-9_4

Vibration tolerance in facilities for advanced technology research and manufacturing should be considered early in the conceptual design phases of a project, and remain in focus throughout the design and construction phase.[1] This process applies not only to new buildings, but also to renovation projects and service or process modifications. The process begins with site vibration evaluations in relation to the relevant criteria. The structural design of the building, the layout of the process and mechanical equipment, and the equipment layout in the laboratory or production spaces are critical to achieving an acceptable level of vibration within the specific areas. Additionally, specifying appropriate means of vibration control, such as vibration isolators, is generally necessary. Finally, attention to vibration when equipment is hooked up is essential.

Considerations for Vibration and Acoustic Noise

The following design parameters should be considered as building criteria are being developed. It is possible that some of them will be deemed unimportant for a particular facility. The design process will be streamlined if the researchers provide manufacturers' site installation guides to the design team at the start of the process. In addition to site environmental requirements, these documents usually provide dimensions and other information important to designers.

Vibration

Vibration can come from a variety of sources. Some internal sources can be controlled through the design and construction of the facility. Internal sources include mechanical equipment, flows through piping and ductwork, people walking, moving materials, and internal transport devices. In some instances, the research or production equipment itself can create vibration. Other sources outside of a facility also must be accommodated. External vibrations include general ambient "micro-seismic" activity due to many sources, near and far. These include road and rail traffic, aircraft, industrial facilities, and construction. Vibration affects different processes in different ways. In some cases, it can cause differential motion within instruments, distorting images. In others, particularly in biological applications, it can degrade a researcher's ability to conduct an experimental process, such as the insertion of a probe into a cell. Vibration can also lead to misalignment in

---

[1] H. Amick, M. Gendreau, T. Busch, and C. Gordon, "Evolving Criteria for Research Facilities: Vibration," *Proceedings of SPIE Conference 5933: Buildings for Nanoscale Research and Beyond* (2005).

processes that occur over extended periods of time, such as time-lapse imaging or photolithography.

Each instrument within a facility could potentially have its own vibration requirements, and each set of requirements could be stated differently. In many cases, the manufacturers' criteria will be provided in a site installation guide. It may be necessary to compile the criteria and convert them to a common format so that the vibration sensitivity of the facility, or portions of it, can be defined.

Generic vibration criteria provide a means to effectively communicate the degree of conservatism required for a facility, or a portion of it. The notion of generic criteria is discussed at length by Amick,[2] along with a discussion of the various units that can be used for expressing vibration criteria. One set of vibration criteria is presented in *IEST RP-CC012.2* "Considerations in Cleanroom Design."[3] Another set of criteria that is popular with researchers working at nanometer scale is that used by the National Institute for Standards and Technology (NIST) for its Advanced Measurement Laboratory, commonly known as NIST-A and NIST-A1.[4]

Virtually all facilities with highly vibration-sensitive equipment will actually have a range of environmental requirements from highly sensitive to not sensitive. To some extent, the cost impact may be minimized by careful examination of the needs of a facility and by grouping areas based upon their vibration sensitivity. There may be considerable savings arising from placing all the highly sensitive instruments in one area of the facility, well away from vibration sources such as mechanical equipment rooms, and placing less sensitive spaces such as offices or generic laboratories between them. For example, many nanotechnology facilities are designed with clearly identified zones, such as mechanical support (vibration sources), cleanroom support (moderately sensitive), cleanroom lithography (highly sensitive), and imaging or metrology (even more sensitive). Each may have a different ideal structural configuration. The relative placement of these spaces should be given careful consideration because it is unnecessarily costly to design an entire facility to meet the most demanding requirements. Moreover, it is also important not to design to meet the requirements of a specific equipment layout

---

[2] H. Amick, "On Generic Vibration Criteria for Advanced Technology Facilities, with a Tutorial on Vibration Data Representation," *Journal of the Institute of Environmental Sciences* **40**(5), 35–44 (1997).

[3] The Institute of Environmental Sciences and Technology (IEST) has published a Recommended Practice (RP) that explores the factors to consider in the design of cleanroom facilities and provides a framework to establish performance criteria—IEST-RP-CC012.2: Considerations in Cleanroom Design. The document is organized into two primary sections: planning and design requirements. The planning section helps users develop a utility matrix to establish the equipment and processes to be used in the cleanroom, to determine the manufacturing layout, and to identify relevant contamination control, life safety, and environmental issues. Ergonomics, budget, and schedule projections are also reviewed.

[4] H. Amick, M. Gendreau, T. Busch, and C. Gordon, "Evolving Criteria for Research Facilities I—Vibration," in *Proceedings of SPIE Conference 5933: Buildings for Nanoscale Research and Beyond* (2005).

only, as the layout may change, and an overly specialized design may hamper future efforts to expand.

Acoustics

Airborne sources of vibration (sound) inside a research facility can be caused by the HVAC system, the research instrumentation itself, people talking, or outside sources. Acoustic noise is a factor in any building, because excessive noise can degrade human performance and lead to discomfort. However, some instrumentation is even less tolerant of sound than people are. Even soft whispers can be problematic. Excessive sound generally effects research by generating internal vibrations within instrumentation, resulting in some of the same problems as excessive structural vibration. Additional information on acoustic isolation is provided in Chap. 5. Acoustic sources of disturbances can often be as important or more so than generic building vibrations since the frequencies of the disturbance can be closer to the resonant frequencies of the instrument structures than are the general building disturbances.

Site Evaluation and Selection

Guidelines and procedures for the measurement of vibration on sites and in cleanrooms should be established early in the project with the vibration specialist or consultant determining vibration specifications.[5]

Vibration

Vibrations are measured with a highly sensitive low-frequency, or seismic, sensor placed on a stake driven into the ground, or on a curbstone, or on something that makes good contact with the ground. Measurements are generally taken at several locations within the footprint of the proposed building. Vibrations in the vertical direction are measured at all locations, but horizontal vibrations are sometimes

---

[5] For documentation requirements for a vibration survey, see IEST RP-CC024, ISBN 978-1-877862-24-3; H. Amick "On Generic Vibration Criteria for Advanced Technology Facilities, with a Tutorial on Vibration Data Representation," *Journal of the Institute of Environmental Sciences* XL(5), 35–44 (1997); H. Amick, L. Vitale, and B. Haxton, "Nanotech I: Site Parameters" and "Nanotech II: Case Studies and Trends," *R&D 2007 Laboratory Design Handbook*, pp. 38–45 (November 2006); and H. Amick, M. Gendreau, and T. Xu, "On the Appropriate Timing for Facility Vibration Surveys," Semiconductor Fabtech, No. 25, Cleanroom Section (March 2005).

measured at only a few of these locations. The data is processed in a spectrum analyzer resulting in plots of amplitude versus frequency.

On a greenfield (unbuilt) site, ambient vibrations are measured at a steady state as well as during representative events that might generate vibrations, such as trains or motor vehicles passing by. When evaluating an existing building, it is important to measure vibrations caused by people walking, talking, and in other activities. In an existing building, the existing mechanical system is virtually always the main source of vibration. The vibration consultant can identify individual items of equipment causing excessive vibration. If the building is to be gutted for renovation, the study should be carried out with the mechanical systems turned off, if possible.

Soil conditions can significantly affect site vibration. For example, sites on which the water table is continuous and shallow (close to the ground surface) can pose problems. Site geotechnical or soils survey reports can be helpful in determining the suitability of a site.

Acoustics

The facility's sensitive research is virtually always indoors, so the building shell will usually provide adequate attenuation for outside noise. However, if the outdoor noise levels are quite high, it may be advisable to conduct a site noise survey, so that the adequacy of the shell's attenuation can be analyzed and enhanced, if necessary, during design. As an example, when the site is near an airport or beneath a flight path, the aircraft noise levels may be high enough that interior noise, even in areas without windows, might exceed stringent acoustical criteria.

The noise should be measured and reported using octave bands or one-third octave bands, rather than single-value representations or decibels. The analyzer settings should be appropriate for interior measurement of the research space of concern, such as maximum-hold, fast response versus slow, and statistical representation or averaging.[6]

---

[6] H. Amick, M. Gendreau, and Y. Wongprasert, "Centile Spectra, Measurement Times, and Statistics of Ground Vibration," *Proceedings of the Second International Symposium on Environmental Vibrations: Prediction, Monitoring, Mitigation and Evaluation* (September 2005).

Design Considerations

Facility Concept and Layout

Adjacency of mechanical equipment areas to vibration-sensitive areas is a sticky issue, but not without precedent in the design of advanced technology facilities. In many cases, the stringent environmental demands of state-of-the-art laboratories require that the mechanical systems responsible for producing close-tolerance environmental control be in close proximity. On the other hand, it is logical to argue that mechanical systems, such as air handling equipment, should be placed as far as possible from vibration-sensitive areas. Adjacent nearby roads and vehicular traffic are also a concern.

Adjacency issues may be dealt with in two primary ways. Where possible, the mechanical systems and vehicular traffic should be located as far as possible from the laboratories of concern. In addition to the obvious benefit of distance for vibration control, there is an associated acoustical benefit since long air ducts can attenuate sound. In cases where systems must be placed close to laboratories, a conservative approach should be taken towards vibration isolation of air handlers and ducts. In general, these provisions are consistent with current state-of-the-art semiconductor facilities where close proximity of mechanical systems is common.

The importance of future flexibility varies from one facility to another. In cases where adaptability is important, increasingly sensitive instrumentation—including instrumentation with increased sensitivity to environmental contaminants, vibration, and noise—will require even more stringent environmental control. These stringent requirements should be addressed by providing for additional control and adequate space.

Foundation

The concept and general layout of a facility greatly determines its vibration levels, as well as the cost to achieve them. For example, vibration control is usually easier to achieve with slab-on-grade floors than with those suspended floors supported on columns. The stiffness or rigidity of a floor is paramount to determining its ability to absorb impacts or shocks without propagating vibration.[7] Stiff floors are preferable in this regard.

Several floor and structural framing schemes are currently being used for vibration control:

---

[7] H. Amick and A. Bayat, "Dynamics of Stiff Floors for Advanced Technology Facilities," *Proceedings of 12th ASCE Engineering Mechanics Conference*, pp. 318–321 (May 1998).

- **Slab-on-grade (NIST Type-A):** A "baseline" floor concept that controls vibration at reasonable cost.

  - **Criterion for Type-A:** Root Mean Square (RMS) amplitude displacement of 0.025 micrometer (μm) at frequencies between 1 and 20 Hz; RMS velocity amplitude of 3 μm/s at frequencies between 20 and 100 Hz.

- **Slab supported by air springs (NIST Type-A1):** Composed of a large block of concrete supported on air springs that are sized to produce a resonance frequency between 0.5 and 3 Hz, the slab vibrations are attenuated at frequencies exceeding 1.4 times the resonance frequency. In most cases, a second "walk-on" floor is provided above the floating slab.

  - **Criterion for Type-AI:** RMS velocity amplitude of 3 μm/s at frequencies below 4 Hz; RMS velocity amplitude of 0.75 μm/s at frequencies between 4 and 100 Hz.

- **Stiff suspended floor:** Used to support cleanroom areas, it is a deep concrete waffle slab supported on columns. Columns are spaced relatively close together to allow a basement but provide high floor stiffness. Lateral stiffness is provided by shear walls.
- **Concrete waffle slab or pan joist:** Used where significant floor stiffness and damping are desirable, such as mechanical rooms; can also be used in areas with minimal, or no, sensitivity to vibration.
- **Steel frame with concrete composite slab:** Used in areas with only minimal, or no, sensitivity to vibration.

Two of these floor types, slab-on-grade and slab supported by air springs, are mainstays of a low-vibration laboratory.

Slab on Grade (NIST Type-A)

A well-crafted slab-on-grade in a quiet location is generally considered to provide the best vibration-controlled environment. The stiffness of the slab, a plate, is supplemented by the distributed stiffness of the soil beneath it. In general, resonance behavior is not observed in a slab-on-grade because of the high damping contributed by the soil.

The primary variables governing the vibration control provided by a slab are the stiffness of the sub-grade (sometimes quantified using the modulus of sub-grade reaction), the slab thickness, and the horizontal dimensions of the slab. Damping properties of the concrete itself are important when considering the propagation of vibrations at higher frequencies (say, those above 30 Hz) that are associated with the impact loads created by hard objects such as tools, gas bottles, or hard heels.[8]

---

[8] H. Amick and P.J.M. Monteiro, "Modification of Concrete Damping Properties for Vibration Control in Technology Facilities," *Proceedings of SPIE Conference 5933: Buildings for Nano-scale Research and Beyond* (2005).

Laboratory slabs should be supported on granular material, the composition and thickness determined by a geotechnical engineer. Slab thickness in the range of 200–300 mm is typical. The sub-base is engineered fill with compaction of 95 % or greater. The slab should be supported with soil of uniform stiffness. If the top of the sub-grade layer is sloping, it is not desirable to have load bearing on both virgin soil and fill, since this might create non-uniform stiffness. In addition to uniform stiffness, care must be taken to avoid the risk of even a slight differential settlement with dissimilar materials. Differential settlement beneath a slab can create voids, which make the unsupported area of the floor slab behave as a suspended floor, with a resonance frequency and without the damping provided by the soil. In other words, the slab would no longer perform as a slab-on-grade.

Conceptually, when a column passes through a floor supporting vibration-sensitive equipment, the intent is to separate the motion of the column from that of the slab. Ideally, there should be an air gap between the two, but typically this is not practical; the gap might form a path or reservoir for dirt and debris. As a compromise, the gap is filled with a foam backer strip and non-hardening elasto-meric caulking on top of the foam. If a column penetrates a slab-on-grade and the foundation is close below, the gap must extend to separate the slab from the footing.

Slab-on-grade floor versus suspended floor: How much is known? The vibra-tion amplitude of walker-generated vibrations on suspended floors (in velocity units) is inversely proportional to the product of stiffness and fundamental reso-nance frequency.[9] In effect, an increase in either or both of the stiffness or resonance frequency, or both, of a suspended floor decreases the velocity amplitude of walker-generated vibrations. Semi-empirical models of this behavior, partially derived from theoretical shock-response equations, have been developed and ver-ified using field data.[10] Given the structural configuration of a floor and the material properties of its components, vibration amplitudes can be predicted with reasonable accuracy.

The literature does not contain similar predictive tools relating vibration ampli-tude to slab and sub-grade properties for a slab-on-grade. It is impossible to define the dynamic properties of the slab, the soil beneath the slab, and the interaction between the slab and the soil with enough accuracy and linearity to develop a predictive model, which in turn could be validated with measured data.

---

[9] E. Ungar and R. White, "Footfall-Induced Vibration of Floors Supporting Sensitive Equipment," *Sound and Vibration* (the Noise and Vibration Control Magazine), p. 10 (1970); E.E. Ungar, J.A. Zapfe, and J.D. Kemp, "Predicting Footfall-Induced Vibrations of Floors," *Sound and Vibration*, p. 16 (November, 2004).

[10] H. Amick and A. Bayat, "Dynamics of Stiff Floors for Advanced Technology Facilities," *Proceedings of 12th ASCE Engineering Mechanics Conference*, pp 318–321 (1998); H. Amick, M. Gendreau, and C.G. Colin, "Vibrations of Raised Access Floors," *Proceedings of the First Pan-American/Iberian Meeting on Acoustics*: 144th Meeting of the Acoustical Society of Amer-ican (December 2002).

A set of relationships, shown below, can be derived that indicates the ratio of increase in stiffness due to a change in slab thickness for a given soil condition:

$$\text{Point Stiffness} = \text{STF} = 8\sqrt{(Dk)} ,$$

k = sub-grade modulus
D = plate rigidity = $E\, t^3/[12\,(1 - v^2)]$
E = Young's modulus
t = slab thickness
v = Poisson's ratio of slab

Logic dictates that the vibration amplitude must be inversely related to stiffness, but a predictive equation, either theoretical or empirical, does not exist at this time. Qualitative discussions must be used to assess the benefits of increased slab thickness.[11]

Based on observation, it seems that slabs-on-grade perform several functions with regard to vibration control.[12] Unfortunately, not enough research has been carried out to define engineering-quality predictive analytical relationships. One of these functions, the dynamic properties of the slab, relates most closely to applied loads such as those caused by people's activities such as walking. The other two, the soil beneath the slab and the interaction between the slab and the soil, relate to propagation of vibration from outside the slab, generally from ambient site vibrations. These functions can be discussed separately, but in reality, all the functions act together.

Averaging of surface wave motion Wavelength of ground vibration (the distance between successive peaks or successive troughs) varies inversely with frequency. The wavelength of a vibration in soil is a function of the density, water content, and wave speed in the soil. High-frequency vibrations have a shorter wavelength than those of low frequency. Imagine a slab of horizontal dimension x. A phenomenon has been observed in which this slab tends to "average" the amplitude of soil vibrations with wavelengths of x or less. Stated another way, the slab acts as a low-pass filter attenuating (reducing, but not eliminating) the amplitudes of vibration components having wavelengths of x or less. This also affects vibrations that propagate in directions perpendicular to the edge of dimension x. A large slab more effectively attenuates low-frequency vibrations than a small one.

This phenomenon also arises when using a small cutout of dimension y in a large slab of dimension x. When vibrations are measured on the cutout slab, the soil

---

[11] H. Amick, S. Hardash, P. Gillett, and R.J. Reaveley, "Design of Stiff, Low-Vibration Floor Structures," *Proceedings of International Society for Optical Engineering* (SPIE) **1619**, 180–191 (1991).

[12] H. Amick, N. Wongprasert, J. Montgomery, P. Haswell, and D. Lynch, "An Experimental Study of Vibration Attenuation Performance of Several on-grade Slab Configurations," *Proceedings of SPIE Conference 5933: Buildings for Nanoscale Research and Beyond* (July 2005).

vibrations at frequencies having wavelengths between x and y are of greater amplitude than those measured on the surrounding solid slab (or than would have existed on the slab prior to the cutout). It was observed on a project that the base vibrations of an electron microscope were made far worse when a saw-cut was made in the slab around the footprint of the microscope.

Vertical stiffening at a point of load application   When a point load is applied to a slab-on-grade, the effective stiffness (the deflection at that point divided by the applied load) is a function of both the slab stiffness alone and the sub-grade stiffness alone. The resulting mechanism might be called a slab-on-elastic-foundation, which has properties similar to a beam-on-elastic-foundation. A qualitative description of this action suggests an axis-symmetric volume of soil beneath the point of application, which contributes vertical resistance to an applied load. The greater the volume of soil mobilized, the stiffer the resultant soil surface. The diameter of this volume of soil is a function of the rigidity of the slab (a function of its thickness to the third power). As shown above, the point stiffness of a slab-on-grade is proportional to $t^{1.5}$, where t is slab thickness.

Stiffening of surface to resist propagation of surface waves   In the propagation of surface waves in slab-on-grade, the motion of a particle is described as a retrograde ellipse. The motion is a combination of displacements in the vertical direction and the direction of propagation. One might envision the vertical component creating a bending wave in a membrane or beam lying on the surface of the soil.

The bending resistance provided by a bending element (plate or beam) on the surface of the soil reduces, to some extent, the amplitude of propagating surface waves. The extent to which the structural element contributes to this resistance depends on the surface wave's wavelength and amplitude. At longer wavelengths (lower frequencies) the curvature is less, so the contribution of the bending element is smaller. Thus, the structural element on the surface tends to be more effective at higher frequencies. Stiffness must be quite high to see an effect at very low frequencies (say, less than 20 Hz).

Long-beam alignment problems such as tilt and attitude variation   The surface wave creates a curved (more-or-less sinusoidal) surface pattern, like ripples in a ripple tank. Therefore, a long-beam apparatus can experience two problems: (1) beam support points can move vertically with respect to each other, causing dynamic misalignment; and (2) beam support mechanisms act as cantilevers and rotate to stay perpendicular to the point on the surface at which they are attached. This latter case also causes dynamic misalignment, in which the amplitude of the misalignment is a function of the spacing between supports.

Joints in slab-on-grade   Joints form definite discontinuities in the slab, serving several functions. Their primary positive effect is that they block vibrations from being transmitted through the concrete itself. Generally, these are at higher frequencies; vibrations propagating through the soil are not affected. However, there are two primary drawbacks to joints. First, they exhibit an edge effect, in which the area near the edge of a slab has a lower vertical stiffness than internal areas of a slab. This affects the ability of the slab to resist dynamic loading. Second, there is an attenuation benefit derived from the horizontal extent of the slab. If a joint creates too small an area of slab, the vibrations could be more severe on the isolated slab.

When discussing joints in a slab, it is important to make a distinction between the two types of joints that can be encountered (from the point of view of mechanics). An isolation joint, or one with a measurable gap of 12 mm, acts as to discontinue the slab with respect to the slab's ability to resist a load or internally transmit a vibration. In general, it does not diminish deflections of the soil beneath the slab, nor does it attenuate ground-borne vibrations. A construction joint, or one with no measurable gap and which exists only to accommodate interruptions in the construction process, does not act as a discontinuity over the range of amplitudes and frequencies of concern in a laboratory. These joints can generally be ignored in the formulation of an engineering model, although they should be constructed to achieve as good a bond as possible between the two concrete surfaces. In subsequent discussion, use of the term "joint" is synonymous with "isolation joint."

In general, expansion joints are not acceptable, except at the perimeter. However, it may not be practical to construct the slab without them. When used, expansion joints should be located so that the enclosed area is as large as practical.

The width of a joint is on the order of 12 mm or more, and it penetrates through the entire depth of the slab. Ideally, the joint should be left open, but it would likely collect debris, which would tend to short-circuit its vibration isolation function. Because of this, it is common practice to use a filler material. It is best to remove forming material and use a soft foam backer strip with a bead of non-hardening resilient grout on top. (Silicone or something with similar properties works well.) If the forming material must remain in place, a composition filler board that is non-healing might be used. These boards come in both forms. A self-healing board creeps plastically to fill in the gap that is left as the concrete shrinks. This gap is a desirable part of the joint mechanism. In fact, the joint is more effective once the gap is formed.

As stated above, a slab-on-grade of large extent generally behaves according to the theory of a slab-on-elastic-foundation. However, near the edges, there is a decrease in stiffness due to the lack of continuity of the slab. Measurements show that the point stiffness (force divided by the resulting deflection) at the edge of a slab of uniform thickness is about one-half that at some distance away from the edge. (In these measurements, it was found that normal stiffness was achieved at a distance of 1–2 m from the edge.) At a corner, where two edges intersect at right angles, the measurements indicated that the stiffness is about one-fourth that at a distance from the edge. Thus, within 1–2 m on either side of a joint in a slab of uniform thickness, the slab is less able to resist the forces generated by a walker or other dynamic load. It also appears, but has not been conclusively proven, that the vibration amplitudes might be slightly higher near a joint. This might be due to minute settlements near the joint (since it is a free edge, it can concentrate stresses), in which case the edge could act as a cantilever and amplify vibrations. Generally, it is recommended that vibration-sensitive equipment not be placed within a meter of a joint.

Slab Supported by Air Springs (NIST Type-A1)

A1 floors are intended to create environments meeting more stringent vibration requirements than the site itself can provide, even with mitigation of the vibrations induced by existing mechanical systems. In concept, the A1 floor utilizes a large inertial mass supported by air springs with low resonance frequencies.[13] An A1 floor is somewhat similar in concept to a conventional optical table on air spring legs (such as a Newport table[14]), only at a much larger scale.

The A1 floor has many of the same drawbacks of an optical table, some of which are enumerated below:

- The support system is soft, so that an applied load (whether due to the placement of equipment or a person stepping on the slab) causes a relatively large displacement; the inertia mass is much larger than a typical optical table top, the springs are stiffer, and the A1 surface less compliant. However, the motion is observable. Excursion of the floor can be kept within prescribed limits using travel limiters.
- The system is susceptible to small excitation forces that might not typically be of concern. For example, the force associated with air flowing from an air conditioning duct could push against an experimental apparatus and move the whole system, causing either offset or oscillation, or both.
- Without extra precautions, an air-spring-supported system is susceptible to rotational motion. Very little resistance is offered against motions about the two horizontal axes or the vertical axis. This characteristic is a problem for some types of experiments and not a problem for others.
- Vibration isolation is provided by means of a more or less rigid mass on a set of springs. There are many resonance frequencies associated with this system, any of which may cause amplification of the base vibrations that excite it. To some extent, the amplification associated with these resonances can be controlled through damping, either naturally occurring damping in the air spring, damping associated with flow from an orifice, or damping induced by active vibration control. However, A1 systems cannot be expected to provide vibration

---

[13] H. Amick and P.J.M. Monteiro, "Vibration Control Using Large Pneumatic Isolation Systems with Damped Concrete Inertia Masses," 7th International Conference on Motion and Vibration Control, Paper 118 (2004); H. Amick, B. Sennewald, N. C. Pardue, E.C. Teague, and B. Scace, "Analytical/Experimental Study of Vibration of a Room-Sized Airspring-Supported Slab," *Noise Control Engineering Journal* **46(2)**, 39–47 (1998).

[14] Newport Corp. http://www.newport.com/Vibration-Control/5681762/1033/section.aspx.

environments as stringent as Type A spaces at frequencies below 10 Hz. Typically, the RMS vibration velocity (measured in one-third octave bands) should not exceed 1 μm/s at frequencies greater than 1.4 times the air spring resonance frequency.

Most experimental work is sensitive to tilt or changes of orientation in space. It is often necessary to control the position of the isolated mass with some sort of active control of position, depending upon the particular application. The quantity and load capacity of airs springs for a particular slab depends upon the weight and shape of the slab.

Only vibration-sensitive equipment is supported on the A1 slab. Personnel and support equipment are supported on a separate floor suspended above the A1 floor, commonly referred to as the walk-on floor. The dynamic loads from personnel activities, including dropped objects, and support equipment is carried along a separate path to the foundation. Vibrational energy traveling back to the A1 floor must first pass through the pneumatic vibration isolation system.

Architectural and Structural Issues

Interior Walls

The structure should be designed to direct vibrational energy through columns into the ground, although it is impossible to completely prevent interior walls from becoming secondary paths of vibration transmission. In conventional construction, walls are attached rigidly to the floor. Rigid connections, however, are undesirable in vibration-sensitive areas, including vibration-sensitive metrology spaces. A wall that is attached to the floor with a rigid connection can transmit its vibrations into the floor, short-circuiting the preferred vibration path and allowing vibrations to travel across a joint in a slab. Demountable partitions should be specified for most walls, with the bottom track of the wall mounted on a resilient connection. Minimizing issues surrounding unwanted secondary paths of vibrational energy transmission are discussed below:

- **Lateral stiffness and out-of-plane vibrations:** Soft walls are subject to dynamic loads applied normal to the surface. These loads include impacts from one side of the wall or the other, or air pressure fluctuations associated with low-frequency sound. The latter is a chronic problem in commercial cleanrooms, and it is necessary to keep tables supporting microscopes or other vibration-sensitive equipment away from walls. Equipment that is bulkhead-mounted through the wall is separated from the wall and does not touch it. Masonry walls provide the greatest resistance to normal loading; demountable partitions provide the least resistance.
- **Connections between wall and floor:** Demountable and drywall partitions can be provided with at least some degree of separation from the floor. It is more

difficult, but not impossible, to provide separation from the floor for a masonry wall. However, if a masonry wall were to be completely isolated, the isolated connection would preclude direct contact between the mortar and the concrete floor.

- **Vibration and noise generated by renovation:** Renovation of any of these constructions will generate vibrations, but demountable partitions probably generate the fewest. The effect of renovation on the other types of construction depends on the procedures that are used and the care taken during demolition and renovation. If the base framework provides a complete separation between the wall and the floor, the vibrations generated in the slab should be much less than in the case of a rigid or resilient attachment.

Interior Sources of Vibration and Noise

There are myriad sources of vibration once a facility is in operation. Air handling devices, pumps, ducting, and piping are part of the technology of any building. These are under the purview of the facility designer. However, there are also sources that are not addressed by the building designer, because they are put in place after the facility is complete.

Gendreau and Amick[15] discuss maturation, which generally involves mechanical equipment such as dry pumps, installed in connection with individual instruments. Over the life of a facility, these sources can have a deleterious effect. It is not economical to design a building to prevent their effects; rather, it is incumbent upon the scientists and facility managers to be vigilant and provide appropriate vibration isolation measures.

Mechanical Systems

Air handling systems are the source of both vibration and acoustic noise. A design strategy to control vibration and noise must be carried through the entire design process, including carefully selecting hardware. Additionally, closely monitoring the installation of vibration control components is essential to their successful implementation.

The following are the primary causative mechanisms in fan-generated vibrations:

- **Motor out-of-balance forces:** The motor generates vibrations at a single frequency (the shaft frequency), which is equal to the motor rpm (the shaft rate),

[15] M. Gendreau and H. Amick, "Maturation of the Vibration Environment in Advanced Technology Facilities," *J. Institute of Environmental Sciences and Technology* **48(1)**, 83–93 (2005).

divided by 60. These vibrations generally occur at slightly less than 1200, 1800, and 3600 rpm (20, 30, and 60 Hz) and appear on a narrowband spectrum as a spike at the shaft frequency. The amplitude of the building vibration that is generated by a motor is typically a function of motor horsepower.

- **Fan out-of-balance forces:** The fan generates vibrations at a single frequency (the shaft frequency), which is equal to the fan rpm (the shaft rate), divided by 60. In direct-drive systems, the fan and motor vibrations are at the same frequency. In belted systems, the fan shaft rate is usually less than the motor shaft rate. These vibrations also appear on a narrowband spectrum as a spike at the fan shaft frequency. The amplitude of the building vibration that is generated by a fan is typically a function of motor horsepower, but it can be dramatically reduced at very low fan shaft rates.

- **Random (broadband) forces due to air turbulence:** Fans are usually contained inside rectangular, box-like enclosures. Turbulence is generated as moving air encounters the inside surfaces of these enclosures and is forced to change direction. This turbulence generates random forces which act against the inside of the enclosure. If a fan housing is rigidly mounted to the floor, these forces are transmitted into the floor structure, and from there to the ground. The forces perpendicular to the floor may excite floor resonances; in this case, the floor may act as a filter and give a predominant frequency or frequencies to the vibrations that propagate into the rest of the structure. Unless the vibrations are filtered by the structure, the building and ground vibrations generated by fan housing turbulence alone are represented by a smooth, haystack-shaped spectrum without spikes.

- **Maintenance and operations problems:** There are many types of vibration problems that can arise through maintenance and operations. For instance, a fan can stall under certain operating conditions in smart buildings in which fan rpm responds automatically to heating or cooling loads. This issue, and its mitigation by means of control limits, must be addressed if variable frequency drive systems are to be used in automatically controlled air handling systems.

In addition, random forces are generated in large ducts by discontinuities in airflow caused by components such as elbows, tees, dampers, and air valves. The spectrum of the forces generated by duct turbulence is like a haystack. Empirical evidence suggests that the amplitude of the dynamic forces is a function of air flow velocity in the duct (higher velocity produces larger forces) and the predominant frequency (at the peak of the haystack) is a function of the inverse of the duct diameter, i.e., a larger duct produces a lower predominant frequency.

The effect of broadband loading from fan housings and duct turbulence tends to be shaped by resonances in the building structural system. It appears that the ground attenuates these vibrations in a frequency-dependent manner but does not shape the spectrum. The effect of turbulence-induced broadband forces to a suspended floor is inversely proportional to the mid-bay stiffness of the floor.

Control of Vibration from Mechanical Systems

The fundamental natural frequency of a piece of equipment mounted on resilient materials that are used to provide vibration isolation (typically spring or neoprene isolators) is given by the following equation:

$$f_n = \left(\frac{1}{2\pi}\right)\sqrt{g/d} = 0.498\sqrt{1/d}$$

where f_n is the natural frequency of the isolation system in Hz; g is the acceleration of gravity, 9.8 m/s^2; and d is the static deflection of the resilient element in meters.

For appropriate isolation of equipment vibration, the natural frequency of the isolation device generally is selected to be less than the frequency corresponding to the drive speed (in Hz) of the supported equipment by a factor of 6 or more, and almost never by a factor of less than 3. Practical considerations, such as the magnitude of the unbalanced forces in a particular piece of equipment, the direction in which the vibratory forces act, practical limitations on spring deflections, and stability of the supported equipment typically result in selection of isolation devices for which the aforementioned factor is substantially greater than 6. This may also occur in instances in which a greater degree of conservatism is deemed appropriate. However, in some cases the factor may be reduced due to practical considerations. For all equipment, vibration isolation requirements are specified by the equipment manufacturers.

Equipment on the roof of the building, such as exhaust fans, is supported above the roof on a steel grillage or dunnage, which transfers the loads to the building columns or major roof beams. The columns and major beams are very stiff and have minimal response to any vibrations transmitted through the isolation systems.

Fans need be specified to achieve stringent dynamic balance requirements. To meet these criteria, the balance must be checked in the factory, before shipping, and confirmed in the field after installation. The final check of the balance in the field should be done with the fans supported on their isolation systems, after the isolation systems have been properly set up and adjusted.

Direct-drive fans may be planned for a variety of reasons, and variable frequency drives can provide the fan capacity as needed. This offers benefits in noise and vibration control. The lack of belt drives helps reduce the broadband vibration of the fans that would otherwise be transmitted through the isolation systems more readily. This also avoids problems over time with maintaining the belts in proper operating tension. The lack of inlet vanes helps to reduce the noise and vibration that the fans create. Speed control also ensures that the fans are operating at the slowest speeds consistent with delivery of the required airflow, ensuring the lowest possible noise and vibration levels. There may be a few fans for which a direct drive is not available. Their use is not expected to have significant impact on vibration or noise.

Any cooling towers should be individually vibration-isolated. The separately supported towers are interconnected using flexible pipe connections. Piping systems are vibration-isolated everywhere unless (1) they are at a great distance from vibration-sensitive areas, or (2) they are quite small in diameter. The piping associated with the pumps should not be suspended from the ceiling, but rather from supplemental members that transfer the loads (and hence the vibration) directly to either the columns or the floor below. When supported in this manner, piping is hung from the supplemental members by isolators, but the supplemental members are rigidly supported from the column or floor. Flexible pipe connections are provided in pipe connections to all vibration-isolated pieces of equipment.

To avoid the transmission of airflow turbulence-induced vibration to the building structure, most of the major ductwork must be resiliently supported. Ducts that may generate or transmit turbulence-induced vibrations should be identified. These duct vibrations—which occur at frequencies of less than 125 Hz and are a significant concern in very large ducts—can be controlled by addressing duct size, length, shape, and layout. This includes transitions and changes in direction, the duct's location with respect to vibration-sensitive areas, flow velocity, and vibration-isolation hardware. All ducts of concern with respect to turbulence-induced vibrations are vibration-isolated.

No structural design will perform well with a poor mechanical design. Furthermore, it is not cost-effective to provide an excessively stout structural design with the intent that the mechanical design will be less relevant. The most cost-effective and historically successful approach is to provide a balanced structural and mechanical design. From the standpoint of vibration impact, the best design adequately separates vibration-generating equipment from vibration-sensitive areas. The layout of the mechanical, electrical, and plumbing (MEP) systems is particularly important and addressed specifically in Chap. 6. As it is used here, "adequately separate" means that enough natural attenuation of the vibration forces is provided by radiation and material damping (in the soil or building structures), in the horizontal and vertical paths between the source and the receiver. This type of vibration reduction is separate from that provided by vibration-isolation hardware, assuming that these are not required for adequately separated sources. Using this design approach, isolation hardware, as discussed below, may be considered a compromise that effectively reduces the required separation distance, but this approach creates increased maintenance and amplification of vibration at the isolator resonance frequency, when this is relevant. Of course, long separation distances are not always practical due to the increased cost in building footprint, piping, thermal losses, and so on. So, the use of isolation hardware may be a desirable compromise.

When equipment cannot be located an adequate distance from sensitive areas, vibration-isolation hardware can reduce the transmission of vibration into the structure at selected frequencies. Transmitted vibration can also be radiated as noise from machine and architectural components, and vibration isolation can also reduce this effect. There are many types of isolation hardware, and their short-term and long-term effectiveness and maintenance requirements can vary dramatically.

In the simplest terms, vibration isolators work by superimposing the response of a single degree-of-freedom spring-mass-damper system on the vibration or force spectrum of a source (machine, pipe, and so on). At frequencies above the resonance of the spring-mass-damper system, the dynamic forces transmitted to the structure are reduced.

Control of Vibration from Electrical Systems

Transformers generate vibration at 120 Hz and multiples of 120 Hz. Unless they are internally vibration-isolated, transformers must be externally vibration-isolated. When transformers are coupled to switchgear cabinets, both need to be isolated.

Any generator must be situated in an area of the generator room where the support is slab-on-grade construction and is vibration-isolated with a two-stage isolation system. The first stage of isolation directly supports the generator and consists of spring isolators that are attached to the base with height-saving brackets. These brackets are positioned outboard of the standard generator steel frame, or rail, base. Positioning the isolators outboard of the generator base reduces the height of the assembly, lowers the center of gravity, and provides a wider footprint for increased stability. The first stage of isolation sits on top of the second stage of isolation, which consists of an inertial mass of concrete with a weight of about twice that of the generator. The inertial mass is supported by air-spring isolators attached to the floor of the generator room. The load is transferred to the ground through suitable foundations. Because the thickness of the inertial mass is substantial and the isolators are tall, the inertial mass is suspended in a pit to keep the top of the concrete mass approximately level with the floor of the generator room. A substantial clear gap is maintained all around and under the inertial mass to avoid contact with the sides and bottom of the pit. (The thickness of the air gap must be sized to avoid coupling between the lab and the bottom of the pit due to stiffness of that air gap.) Steel structural members can be embedded in the concrete mass and project to the sides of the mass into pockets in the sides of the pit. These steel members' hanging points are located along the centerline of the isolators, which are set back from the edge of the pit. The suspension rods transfer the load of the entire assembly to structural members that, in turn, transfer the loads to the tops of the isolators.

Acknowledgement Thanks to Hal Amick, who integrated material from various sources into the initial draft.

Bibliography

J. Ambrose, J.E. Ollswang, *Simplified Design for Building Sound Control* (Wiley, New York, 1995). ISBN 0 471 56908 9

H. Amick (ed.), *Buildings for Nanoscale Research and Beyond* (proceedings volume), Proceedings of SPIE 5933 (2005)

H. Benaroya, *Mechanical Vibration: Analysis, Uncertainties, and Control.* Mechanical Engineering, 2nd edn. (Marcel Dekker, New York, 2004). ISBN 0 8247 5380 1

M.J. Crocker, *Handbook of Noise and Vibration Control* (Wiley, Hoboken, NJ, 2007). ISBN 978 0 471 39599 7

C.W. de Silva, *Vibration Damping, Control, and Design.* Mechanical Engineering Series (CRC, Boca Raton, FL, 2007). ISBN 978 1 4200 5321 0

F. Fahy, *Advanced Applications in Acoustics, Noise and Vibration* (Taylor & Francis, Oxford, 2004). ISBN 0 415 23729 7

M. Gendreau, H. Amick, Chapter 29: Micro-vibration and noise, in *Semiconductor Manufacturing Handbook*, ed. by H. Geng (McGraw-Hill, New York, 2005)

M. Hirschorn, *Noise Control Reference Handbook* (Industrial Acoustics, New York, 1982). Available at http://books.google.com/books/about/IAC_Noise_Control_Reference_Hand book.html?id=Xm-sHAAACAAJ

M. Möser, S. Zimmermann, R. Ellis, *Engineering Acoustics: An Introduction to Noise Control* (Springer, Berlin, 2009). ISBN 978 3 540 92722 8

S.D. Snyder, *Active Noise Control Primer* (Springer, New York, 2000). ISBN 0 387 98951 X

E.E. Ungar, H. Amick, J.A. Zapfe, Chapter 13: Vibration considerations for sensitive research and production facilities, in *Structural Dynamics of Electronic and Photonic Systems*, ed. by E. Suhir, D. Steinberg, T.X. Yu (Wiley, Hoboken, NJ, 2011)

I.L. Vér, L.L. Beranek, *Noise and Vibration Control Engineering: Principles and Applications* (Wiley, New York, 2006). ISBN 978 0 471 44942 3

L.F. Yerges, *Sound, Noise, and Vibration Control* (Krieger, Malabar, FL, 1978). ISBN 978-0898746549

Chapter 5
Acoustic Noise

Abstract Research and manufacturing instruments are sensitive to internal vibration that can be excited by the external acoustic environment. The degree to which acoustic excitation occurs depends on many factors, but it happens primarily when there is correspondence between the resonance characteristics of the instrument and the frequency content of the acoustic environment. For high performance measurement and imaging laboratories, a noise criteria of NC 25 is recommended; standard laboratories use NC 40. This chapter addresses the various sources of acoustic noise; the mechanisms by which it can interfere with instruments; an approach to determining sources of noise generated in and propagated by the building; and techniques used to isolate the lab or instrument from these sources.

Introduction

Research and manufacturing instruments are sensitive to internal vibration that can be excited by the external acoustic environment. The degree to which acoustic excitation occurs depends on many factors, but it happens primarily when there is correspondence between the resonance characteristics of the instrument and the frequency content of the acoustic environment. In addition, low-frequency acoustic pressure fluctuations (or infrasound) can interfere with exposed energy beams used in experimental work. Adverse acoustic environments, such as those often found in operating laboratories, can affect the resolution achievable by the instrument.

This section of the report includes:

- A review of the basic terminology and criteria used in the acoustical design of advanced technology facilities;
- Discussion of the mechanisms by which acoustic noise can interfere with instruments;
- An approach to determining sources of noise generated in and propagated by the building, both internal and external to the lab; and

© Springer International Publishing Switzerland 2015
A. Soueid et al. (eds.), *Buildings for Advanced Technology*, Science Policy Reports,
DOI 10.1007/978-3-319-24892-9_5

- An inventory of techniques used to isolate the lab or instrument from these sources.

Acoustics Terminology Review

This chapter focuses on the parameters that are most important in defining room criteria and equipment specifications with respect to acoustic noise. It is limited to acoustic phenomena, the impact of vibration with air as the medium (i.e., pressure fluctuations in air), as opposed to the same phenomena propagated through built structure (which is addressed in Chap. 4). Some traditional examples of acceptable noise levels in various spaces with respect to their human occupants are shown in Table 5.1. In this chapter we will discuss the application of these types of criteria, and expand on them as necessary for instruments in research environments.

Pressure Versus Power, Decibels

Noise is pressure fluctuation in air. Ears and machines respond to these variations. The acoustic pressure that is measured or heard is a function of environmental factors such as surface absorption, reflection and reverberation, as well as the inherent characteristics of the sources of noise. On the other hand, sources of noise are best characterized by their inherent qualities, qualities not dependent on an environment, such as the amount of power they radiate, as a function of frequency, and their specific radiation pattern in space (directivity), and in time. Both pressure and power are commonly expressed in decibels.[1]

Types of Noise

In general, noise can be described as variations in air pressure over time. The fluctuations can occur at specific frequencies, or randomly over a wide range of

---

[1] Due to the large dynamic range (12 orders of magnitude for human hearing), both pressure and power (and other phenomena) are frequently expressed in units of decibels (dB), a logarithmic scale relative to some reference level. A complete statement of a unit of measurement in decibels must be accompanied by a reference level. The full expressions, including the reference units necessary to clarify which phenomenon is numerated, are:

Pressure: dB re 20 μPa
Power: dB re 1 pW

Table 5.1 Noise requirements for occupied spaces

Use/space	Recommended NC[a]
Clean rooms (Class 1–100)[b]	60
Laboratories with fume hoods (sash open)	50
Lab support spaces	50
Lab equipment corridor	50
Light maintenance shops	50
Computer rooms	45
Corridors and public circulation areas	45
Shop classrooms	40
Laboratories without fume hoods[c]	40
Open-plan offices	40
Private offices	35
General classrooms	35
Libraries	30
Executive offices	30
Large lecture rooms	30
Conference rooms	30
Teleconference rooms	25
High-performance and imaging laboratories	25
Auditoria	25

[a]This compilation is based on the following references, and others: ASHRAE HVAC Applications Chapter 42 Sound and Vibration Control (1991); Howard F. Kingsbury "Review and Revision of Room Noise Criteria" Noise Control Engineering Journal 43-3 May-June 1995; American National Standards Institute Criteria for Evaluating Room Noise ANSI S12.2-1995; Robin M. Towne et al. "The Changing Sound of Education" Sound and Vibration January 1997. The noise criterion curves (NC) were developed by Beranek in the 1950s to establish conditions for speech intelligibility and general living environments. They are expressed as a series of curves, which are designated NC-30, NC-35, and so on, according to where the curve crossed the 1750 Hz frequency line in an octave band designation (octave band sound pressure level, dB re 20 μPa

[b]Certain instruments (e.g., e-beams) may require better acoustic performance, which may require acoustic isolation in the cleanroom, and lower or controllable air flow velocities, or both

[c]In laboratories with snorkel exhaust systems, the criterion is assumed to apply with the snorkel closed or powered off

frequencies, or a combination of the two. Thus, noise can be separated into several general classes[2]:

1. Tonal versus random (Fig. 5.1):

    (a) Tonal noise is periodic noise at a single frequency. It is usually associated with periodic processes, such as rotating or reciprocating mechanical equipment, or magnetostriction in transformers.

---

[2] Sections of this description of acoustic phenomena and measurement methodologies are excerpted and adapted from M. Gendreau and H. Amick, "Micro-vibration and Noise Design," Chapter 39 of *Semiconductor Manufacturing Handbook* edited by H. Geng (McGraw Hill, 2005), ISBN 0-07-144559-5. See this reference for further details on noise and vibration requirements for advanced technology facilities.

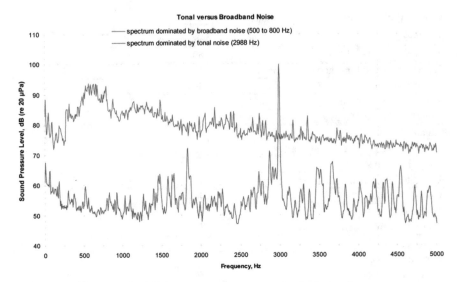

Fig. 5.1 Illustrations of broadband (*top line*) versus tonal (*bottom line*) noise

   (b) Random noise, also referred to as broadband noise, contains random energy at many frequencies. Sources of random noise include turbulent fluid flow, and impacts, such as footfall and door closures.

2. Steady-state versus transient:

   (a) Steady-state[3] noise does not change significantly, on the average, over time. This is produced by continuous or relatively continuous sources, such as operating machinery, constant traffic, and HVAC equipment.

   (b) Transient noise is time dependent. Transient sources include single passing vehicles, impacts, and intermittent machinery such as robotics and automatic material handling systems.

A measurement characterizing the variation in amplitude over time—in representational form known as time domain—is necessary to record all of the information about the environment. However, given the complexity of the environment from the standpoint of frequency, time domain data do not provide a simple readable representation of the significant components of the noise in high-technology facilities. (There are exceptions to this, of course, with regard to acoustic transients.)

Perhaps the most useful aspect of time domain data is that it is completely re-analyzable, and can be used to produce other representations of data. The reverse is not true, since basic time domain data cannot be reconstructed from other data

---

[3] Strictly speaking, continuous random vibration or noise is said to be "stationary," and the term "steady-state" is used for tonal content. However, we will use the latter term interchangeably.

representations, giving the time domain representation the distinction of having the greatest information depth, although often unreadable in its complexity.

Structures and tools are subject to resonance response, meaning that they are more sensitive to vibration and noise at certain frequencies more than at others. With many tools, the sensitivity to noise or vibration will be based upon whether or not internal resonances are excited. This suggests that it is important to have knowledge of the frequency content of the vibration or noise. When noise is represented as a function of frequency, it is said to be in the frequency domain. There are several ways to obtain data in the frequency domain.

Using Fast Fourier Transform (FFT) methods, a snapshot of the time domain data can be transformed into a representation of amplitude as a function of frequency over the measurement period.[4] Definition of the parameters and techniques used in the acquisition of the FFT samples is necessary for accurate representation of the data, and to allow adequate mutability, if needed. These include:

- **Type of noise environment**. Presence of various types of noise may dictate measurement parameter settings:

  - **Stability of the environment during the sample period**. If the environment is steady-state, representations of the average environment may be used, as long as the average time is long enough to represent a stable value. In the case in which the environment is not steady-state, representations of the maximum value obtained at each frequency may be used. If the environment contains significant transients, representations of the average value are usually not useful unless accompanied by time domain data for the sample period. In either case, maximum representations are useful if one wishes to know the maximum impact during the measurement period. It is especially important in this case to state the measurement time.

  - **Tonal versus broadband noise**. The frequency resolution of data may be influenced by the level of detail needed to identify tonal frequencies or resonances. Also, the type of frequency analysis used may be determined by the presence of tonal or broadband noise.

- **Measurement time and methodology**: The measurement time must be long enough to adequately characterize the lowest frequency of the desired frequency range. The measurement methodology typically uses one or more of several standard representations of average or maximum noise levels:

  - Average **using linear or exponential averaging**[5]: L_{eq} (the root mean square (RMS) equivalent energy average),[6] L_n (statistical centile levels)[7]

---

[4] FFT analysis results in spectra with amplitudes spaced at uniform increments of frequency, which we call the spectrum's *resolution*.

[5] Both averaging methods consider all data collected over the measurement period; the former method (typically preferred for purposes described in this report) considers each moment of equal importance, the latter method exponentially weights the most recent data as more important.

[6] The equivalent-continuous sound level, L_{eq}, is the level of steady (non-varying) sound which, for the measurement period, has the same sound energy as the time-varying sound.

[7] A statistical centile level (L_n) is the level that is exceeded n percent of the measurement time.

– Maximum: L_{max} (maximum RMS),[8] L_{peak} (greatest instantaneous sound pressure)

Depending on the nature of the noise, each of these averaging methodologies may produce different results, so definition and representation of the appropriate methodology is critical.

- **Time constant**: The RMS averaging time, or the time constant as it is often referred to, may affect the measured amplitude, especially in the presence of transient noise. There are several standard constants (slow, fast, impulse; 1000, 125, and 35 ms respectively). Others may be used when appropriate and clearly documented.
- **Frequency range**: This is the overall range of frequencies used in the analysis.
- **Bandwidth**: The measurement bandwidth affects the time for each individual sample and the frequency resolution of the data. Bandwidth can also present a potential limit to the mutability of the data. Although relatively low resolution data can be constructed from higher resolution data (e.g., transforming constant narrowband data with a resolution of 0.1–1 Hz narrow bandwidths); high resolution data cannot be constructed from lower resolution data. Selecting the appropriate bandwidth is determined by the level of detail needed in frequency resolution.
- **Measurement position**: In noise measurement, the position of the microphone may influence measured results due to spatial anomalies, such as the spatially invariant effect of standing waves, and proximity to sources. Spatial averaging (by slowly traversing the space with the microphone) is often used when representation of the average environment in the room is needed. This might be limited to the proposed position of a tool, for example. However, in measurement of infrasound (low-frequency noise), microphone motion may cause interference, and use of a fixed microphone position is always recommended.

Another form of spectral representation is the proportional bandwidth spectrum, in which the frequency bandwidth of each band is proportional to the center frequency of that band. The two most common forms are the octave band (in which each center frequency is double the previous frequency) and the one-third octave band (in which there are three proportional bands per octave and the bandwidth is 23 % of the band center frequency). These are illustrated in Fig. 5.2.

A third form of data representation used in acoustics is the single-value quantity, such as the A-weighted sound level, commonly denoted with units of dBA. There are several standard weighting curves, the most popular being A and C, which are used to emphasize content at some frequencies and de-emphasize content at others.

---

[8] The L_{max} methodology produces the maximum RMS noise level recorded during the measurement period.

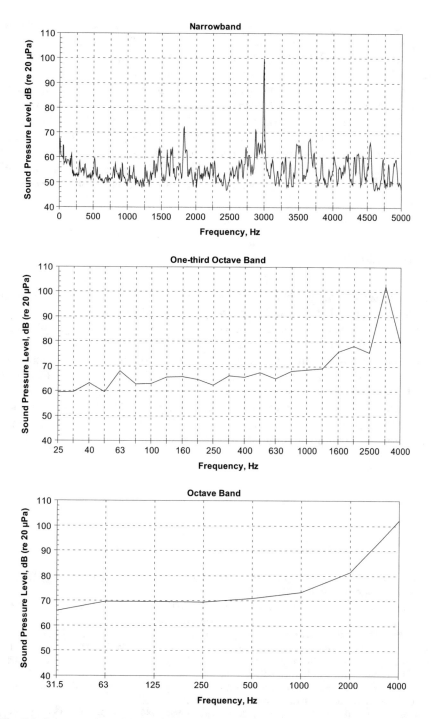

Fig. 5.2 Representation of the same noise data (overall value of 102 dB in the 0–5000 Hz frequency range) in various bandwidths

A-weighting is used to simulate how the human ear perceives sound at moderate amplitudes, de-emphasizing content at low frequencies. The most important shortcoming of A- and C-weighting in measurements pertaining to tools is that it is impossible to extract frequency characteristics from the data.[9]

As we have noted, site and building noise can be classified as steady-state random much of the time. For this reason, and due to the frequency dependence of tool vibration sensitivity, basic analysis and representation of noise environments in advanced technology facilities is usually in the frequency domain, represented as an average (for steady-state noise) or maximum (for transient noise) value. This representation uses constant narrow bandwidth which may later be simplified in wider bandwidth formats for comparison with specific tool specifications or generic criteria.

Noise Criteria

There are several families of noise criterion curves with frequency weighting based on human perception. These are generally considered to be within the range of 20–20,000 Hz and 0–120 dB re 20 μPa (at 1000 Hz; the dynamic range of hearing varies with frequency). The various frequency-based curve families, given their common subject, are similar in shape. However, there are some exceptions and variations, as shown in Fig. 5.3 (25 dB at 1000 Hz) and Fig. 5.4 (60 dB at 1000 Hz).[10] They are variously standardized, described, or recommended by the International Organization for Standardization (ISO), American Society of Heating, Refrigerating and Air Conditioning Engineers (ASHRAE), and other organizations, and varyingly appropriate for use in different situations.

---

[9] M. Gendreau, "Specification of the Effects of Acoustic Noise on Optical Equipment," Noise & Vibration Worldwide, Vol. 32(4), (April 2001).

[10] There are number of sources for the criterion curves cited in Figs. 5.3 and 5.4. The NC curves were described in the footnotes of Table 5.1. The Balanced Noise Criteria (NCB) curves are intended to replace the NC curves and were published as ANSI S12.2-2008. The NR (noise rating) curves were developed by the International Organization for Standardization to determine the acceptable indoor environment for hearing preservation, speech communication and annoyance. NR is commonly used in Europe; NC is more common in the U.S. The PNC (Preferred Noise Criterion) curves are often used to judge the acceptability of ventilation and other background broad-band noise sources; they have more stringent lower frequency requirements than the NC curves. The RC (Room Criteria) were developed around 1980 by Blazier for ASHRAE as a response to low- and high-frequency problems with NC; they are considered spectrally balanced.

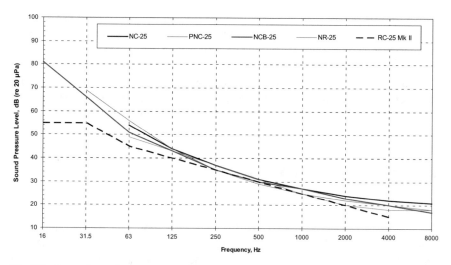

Fig. 5.3 Comparison of perception-based criterion curves (range commonly used for "quiet" labs)

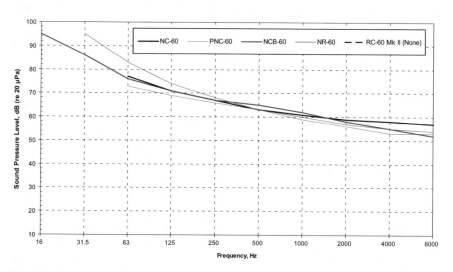

Fig. 5.4 Comparison of perception-based criterion curves (range commonly used for "noisy" labs, e.g., cleanrooms)

Impact of Noise on Research and Process Equipment

The perception-based curves above are often used as design criteria for environments housing people as well as tools. However, these curves do not always accurately reflect actual tool requirements. There are currently no widely accepted generic noise criteria for research tools[11] and, whenever possible, spaces are designed to specific requirements of individual tools. This situation is less than ideal in cases where the specific instruments which will be used have not been identified, or when a space is designed for future unknown requirements.

Understanding some of the mechanisms that make advanced technology equipment sensitive to noise is important, as it helps to determine likely tool specifications; criteria with which to develop lab noise requirements.

Mechanisms by Which Acoustic Noise Interferes with Equipment

The resolution that can be achieved by an optical tool is in part a function of differential vibration between critical elements in the tool; for example, between a lens and the observed target. Vibration of elements within a tool can be stimulated by vibration sources within the tool, external vibration sources, and acoustic noise. Acoustic noise in the environment causes exposed elements of the tool to vibrate. The vibration is then transmitted through the structure of the tool to sensitive internal components. Acoustic noise or pressure fluctuations can also interfere with exposed beams used in tools and research.

The likelihood of impact from acoustic noise, which varies somewhat from tool to tool, depends on the frequency range of the noise. A low frequency range, 2–200 Hz, has a low to medium probability of impact. This is because the acoustic wavelength is significantly longer than the dimensions of the tool structures, and coupling between the two is relatively inefficient. However, exceptionally low frequency pressure fluctuations (10 Hz and lower) may interfere with tools with open beams (some interferometers, atomic force microscopes, etc.), and those supported on air springs, so the impact of infrasound must be considered in certain cases.[12] Noise in the mid frequency range, approximately 200–2000 Hz, has a higher probability of impact for most tools. In this frequency range, especially at coincidence frequencies (where acoustic and structural bending wave speeds are equal) and above, the structure is more likely to be excited by acoustic energy. Finally, in the high frequency range (above 2000–10,000 Hz) the impact probability

---

[11] There are current efforts underway to develop generic tool noise criteria by IEST and others. Also see: M. Gendreau, "Generic noise criterion curves for sensitive equipment," Invited presentation, Proceedings of Acoustics 08 Paris, Paris, France (29 June–4 July 2008).

[12] M. Gendreau and B. Sennewald "Infrasound in laboratories: criteria, phenomena, and design," Proceedings of SPIE 5933: Buildings for Nanoscale Research and Beyond (August 2005). ISBN 9780819459381; M. Gendreau, "The effect of varying acoustic pressure on vibration isolation platforms supported on air springs," Proceedings of the 16th International Congress on Sound and Vibration (ICSV16), Kracόw, Poland (5–9 July 2009); M. Gendreau, "Measurement techniques used to verify the cause and nature of low-frequency noise in rooms," Proceedings of the 16th International Congress on Sound and Vibration (ICSV16), Kracόw, Poland (5–9 July 2009).

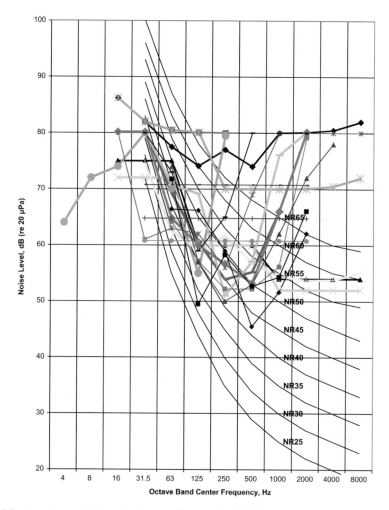

Fig. 5.5 Several manufacturers' noise specifications for SEMs compared with the standard NR curves

is again lower. The acoustic excitation of structures due to noise in this range is less of a concern because there is less acoustical energy available with increasing frequency, among other reasons.

While the quality of information available from manufacturers about the sensitivity of their instruments to noise varies widely, the situation does appear to be improving. Several manufacturers provide detailed tested (as opposed to estimated) specifications for acoustic noise for their tools. For example, Fig. 5.5 contains noise specifications for several models of SEMs of various resolutions, and Fig. 5.6 contains a similar plot for TEMs. These tool specifications have been plotted over the commonly used NR curves. It is clear that the general trend (perhaps best represented by the minimum value at each frequency for all SEMs and TEMs) does

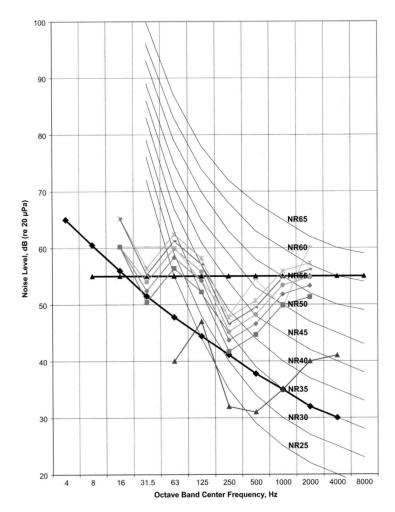

Fig. 5.6 Several manufacturers' noise specifications for TEMs compared with the standard NR curves

not follow the general shape of the NR curves, especially at low and high frequencies. Use of a particular perception-based curve may be conservative at high frequencies, and anti-conservative at low frequencies.

Also, note that some of the tools have frequency requirements that extend lower than the standard curves, into the infrasonic range. This is important to consider. Specific infrasonic performance can dictate the selection of architectural and mechanical systems and should be determined early in the building design process. Design for infrasonic noise control is complicated by the lack of acoustical data below 63 or 125 Hz for mechanical equipment, silencers, absorptive treatments, partition transmission loss, etc.

For these reasons, using the requirements specific to the tools being installed is best. Unfortunately, this is not always possible. For unknown future or generic installations, using a judiciously modified NR or NC curve is recommended or, preferably, a generic curve based on instruments of resolution similar to the lab requirements. Due to the specific sensitivity of research equipment to the frequency of noise, using overall noise referents such as dBA and dBC is not recommended.

Laboratory Noise Sources and Control

The final section of this chapter regards the types of noise sources encountered in an advanced research facility, the dominant frequency ranges of the sources, and an outline of noise control methods. In every case, there are usually many ways to control noise, but it is important to note that the layout of a facility may limit options, either because of physics or cost. This is why it is important to consider the acoustical requirements of the sensitive instrumentation from the outset of the building design process.

Noise sources: Listed below are noise sources that can impact operating laboratories and cleanrooms, and the typical frequency ranges of these sources:

- External noise sources:

  - These include air and ground traffic, and mechanical equipment. The noise is tonal and broadband, and typically ranges from infrasonic to 500 or 1000 Hz. These noise sources may present a low frequency problem, depending on the noise reduction provided by building façades and internal walls.

- Internal noise sources:

  - HVAC fans generate broadband and sometimes tonal noise, usually predominant in the 63–250 Hz octave bands, or up to 1000 Hz with small fans (e.g., as used in fan-filter units (FFUs) and fan wall units). There is sometimes infrasonic noise, and motor or other tonal noise, up to 1000 or 2000 Hz.
  - HVAC diffusers generate flow noise, which is broadband and occasionally tonal in the 1000–8000 Hz range.
  - Tools and support equipment generate broadband and tonal noise in the 250–2000 Hz range. Occasionally, they generate higher frequencies, especially from air flow or air activated components in tools.
  - Fluid flow, valve, and air leakage noise is generally broadband at 1000 Hz and above.

- **Personnel noise**, such as voices, is in the 500–2000 Hz range and is often controllable by the users of the facility.

Noise Control

Fortunately, once the room criteria and nature of the noise sources is known, in many cases standard (and sometimes recording-studio-standard[13]) noise control design can be used. In other cases, less conventional methods must be applied. In any case, attention to detail is required at all stages of the design and construction of the building, including when installing the tools, especially for low-noise environments.

The following are some of the considerations that may be taken in the design of a low-noise laboratory:

HVAC noise can be generated by several elements along the path from the fan to the diffusers (or radiant cooling systems) serving the room. All of these must be considered with respect to the fan operating point and duct air flow velocity. Noise control at the source (fan units) may be by specification, leaving internal air handling unit noise control to the unit manufacturer with the use of dump walls, splitters (internal silencers), devices promoting laminar inlet flow, etc. Along the air flow path, there is the option of using absorptive lining (when allowed) or silencers. Finally, terminal elements (diffusers, duct elements, air terminals, snorkels, fume hoods, etc.) must be selected to be compatible with the room noise requirements and air flow velocity.

Tool support equipment such as dry pumps, chillers, and power conditioners, can be a major source of noise. Contemporary labs often use pump chases that are acoustically separated from the labs. Or, by selecting equipment with lower noise ratings (assuming good data is provided by the manufacturer), local pump enclosures and appropriate absorptive equipment casing materials can be used.

Controlling noise from sources external to the lab is also important. A good facility layout that separates noisy areas (mechanical rooms, duct shafts, toilets, public corridors) from the labs is extremely helpful. Careful attention to the design of the walls, floors, and ceilings is critical, considering the specific nature of the noise sources outside the lab and the lab noise criterion. For walls, either gypsum wallboard or masonry constructions may be used, depending on the circumstances and attenuation frequency range of interest. For example, in some instances, a double-layer, double-stud gypsum wallboard construction is a better option than a single concrete block wall for acoustics. No partition will perform as rated without selecting appropriate doors, windows, etc., and considering the impact of penetrations for utilities and services.

Infrasonic noise control must consider its primary sources: large fan HVAC systems, aircraft, wind and other meteorological phenomena, etc. Special constructions and operating conditions are necessary for low infrasonic noise environments.

Finally, it is often beneficial to provide reverberation control in the most sensitive labs to reduce the noise impact from HVAC, installed tools, support equipment, and personnel in the laboratories. However, this is never as efficient

---

[13] P Newell, Recording Studio Design (Focal Press, 2008), ISBN 978-0-240-52086-5.

as controlling noise at its source. The specific amount and quality of absorptive treatments applied to walls and ceilings depends on the specific circumstances and, when relevant, cleanroom particulate and outgassing considerations.

Acknowledgement Thanks to Michael Gendreau, who composed the initial presentation and figures, and to James Murday, who integrated material from several other sources.

Bibliography

J. Ambrose, J.E. Ollswang, *Simplified Design for Building Sound Control* (Wiley, New York, 1995). ISBN 0-471-56908-9

M.J. Crocker, *Handbook of Noise and Vibration Control* (Wiley, Hoboken, NJ, 2007). ISBN 978-0-471-39599-7

M. Hirschorn, *Noise Control Reference Handbook* (Industrial Acoustics, New York, 1982). http://books.google.com/books/about/IAC_Noise_Control_Reference_Handbook.html?id=Xm-sHAAACAAJ

M. Möser, S. Zimmermann, R. Ellis, *Engineering Acoustics: An Introduction to Noise Control* (Springer, Berlin, 2009). ISBN 978-3-540-92722-8

SAE Institute, *Audio Reference Material*, http://www.sae.edu/reference_material/audio/pages/fullindex.htm

S.D. Snyder, *Active Noise Control Primer* (Springer, New York, 2000). ISBN 0-387-98951-X

I.L. Vér, L.L. Beranek, *Noise and Vibration Control Engineering: Principles and Applications* (Wiley, Hoboken, NJ, 2006). ISBN 978-0-471-44942-3

L. Yerges, *Sound Noise, and Vibration Control* (Van Nostrand Reinhold, New York, 1978)

Chapter 6
Disturbances due to Building Mechanical Systems

Abstract Today's advanced technology buildings require extensive mechanical equipment for building heating, cooling, ventilation, and filtration. The challenge is to design, construct, and maintain an environment that is relatively quiet and relatively free of vibration, both for research personnel, research equipment, and neighboring facilities. Reduced thresholds for both noise and vibration levels from all sources, including systems effects such as flow turbulence in ducts and air inlets have required that every aspect of building mechanical systems be carefully studied and designed to minimize such effects. This chapter addresses the vibration and noise induced by building mechanical systems and presents several cases in which these problems were solved.

Introduction

Today's advanced technology buildings require extensive mechanical equipment for building heating, cooling, ventilation, and filtration. The challenge is to design, construct, and maintain an environment that is relatively quiet and relatively free of vibration, both for research personnel and equipment. Significant challenges include:

- Most research facilities are home to scientists and researchers who spend the majority of their daily life inside the building.
- Research equipment is sensitive to ambient noise and vibration (caused by the operation of mechanical equipment, among other sources) with a threshold that is becoming less tolerant. Vibration limits at reduced scales are currently being specified by the research equipment manufacturers.
- Major mechanical equipment (e.g., cooling towers, air handling units, and pumps) is increasingly relegated to a building's exterior, creating greater potential for noise impact on neighboring facilities.
- Most of today's research facilities are also used for teaching, increasing the need to minimize background noise.
- Zoning noise restrictions can be problematic, as the restrictions typically apply to night operations. Most research facilities operate 24 hours a day, 7 days a week.

© Springer International Publishing Switzerland 2015
A. Soueid et al. (eds.), *Buildings for Advanced Technology*, Science Policy Reports,
DOI 10.1007/978-3-319-24892-9_6

Historically, noise and vibration concerns from building mechanical systems have been related to rotating mechanical components such as pumps, fans, and compressors. Noise and vibration problems resulting from these components typically could be mitigated by sound attenuators, vibration isolators, and acoustical enclosures. However, there are potential noise and vibration problems (also referred to as the system effect) that result from poor aerodynamic design such as duct layouts and use of inferior or inappropriate system components. These aerodynamic generated noise and vibrations are most often invisible during the design process, but are noticeable once the system is up and running.

This section addresses noise and vibration induced by building mechanical systems and presents several cases in which these problems were solved.

Examples

Usually, noise and vibration problems adversely impacts three sensitive receptors: the building occupants (creating uncomfortable interior environment), a facility's neighbors, and the research equipment. The following examples address all three noted sources.

Electron Microscope Operation and Noise Levels

Manufacturer specifications for an electron microscope limit the background noise level in the microscope suite to no greater than 55 dB on any of the standard octave bands frequencies. The client requested that the background noise be reduced 5 dB lower than that of the microscope manufacturer's specifications, due to future plans to purchase a more sophisticated microscope that would require even quieter noise environment.

The building's mechanical installation, as designed, resulted in a noise level that was measured at 59 dB at 31.5 Hz frequency band. Field measurements revealed that air turbulence at the return air duct above the microscope suite was responsible for the 31.5 Hz noise problem. To solve the problem, ductwork was reconfigured to create and maintain a better aerodynamic flow, air velocity was reduced, duct transitions components were replaced with smooth-radius elbows, and volume dampers were moved further away from the room air outlets.

Flow at Fan Inlets

Exhaust fans in a mechanical room generated strong tonal sound that was transmitted to adjoining offices through the mechanical room partition wall structure. The design mechanical duct installation was designed to bring air from lower level

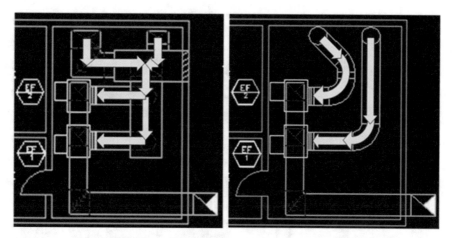

Fig. 6.1 Ductwork layout before and after modifications (courtesy of Ove Arup)

laboratories fume hoods and general exhausts into a common overhead duct connected to two identical exhaust fans. The intention of the exhaust air system was to provide 100 % exhaust air volume using constant revolutions per minute fans. A report on air balance identified several issues: static pressure readings were much lower than that of the design, while the airflow rate was also lower. Following the standard fan laws, the relationship between the measured static, airflow, and fan speed suggested that these fans were likely overcoming much higher static pressure than that shown by the air-balance report, i.e., a system effect. The system effect was likely originating from the non-uniform airflow at the inlet of the fans.

Corrective measures were implemented to reduce the air swirl at the fan inlet; smooth-radius duct elbows were installed (see Fig. 6.1). Laminar airflow into the fan blades lowered the tonal sound to the level that was deemed acceptable to the occupant of the adjoining office space. Air balance readings obtained after the installation of the inlet elbows, indicate the airflow volume, static pressure, and the fan speed values improved greatly, performing much closer to the specified design values.

Swirling Air at Inlet to Vane Axial Fan

Vane axial fans were installed in the exhaust air plenum serving a laboratory building. Each exhaust air plenum was connected to two fans, one fan serving as a standby (see Fig. 6.2). While, the exhaust airflow volume from the laboratories would vary, the fans constant volume was maintained through outside air louvers placed at the end of the plenum controlled by a motorized damper. Each fan had flex connectors at the inlet with the diameter equal to that of the fan inlet. Once operated, the fan closer to the outside air louver exhibited noticeable noise and vibration. Field measurements showed poor airflow (air turbulence) at the inlet of

Fig. 6.2 Typical lab building exhaust air duct layout (courtesy of Ove Arup)

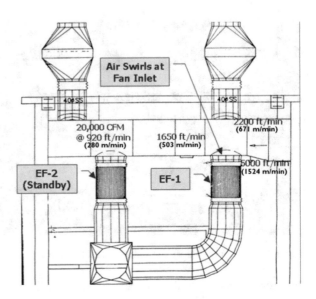

the fans. Typically, axial flow fans operations are influenced by inlet air conditions. Uniform flow to the fan blades is necessary in order for the fans to follow the manufacturer's published performance curve. Here, air was entering the fan from two sides: from the duct facing the fan inlet, and from the outside louver positioned 90° from the fan inlet.

The inlet condition was such that the flex connector, (see Fig. 6.3), was drawn into the fan inlet further restricting air movement, and increasing air velocity from 1200 to 1500 m/min. Sound level readings taken with each exhaust fan operating individually indicated that the fan unit mounted closer to the outside air louver was approximately 10 dB noisier that the fan on the opposite side. Considering the likely operational requirements of the fans, (at some point in the operation cycle of the building these fans will move 100 % of the air coming from the space), it was decided that reducing the fan speed was not a viable alternative. Instead, a bell-mouth duct piece was fabricated and installed at the inlet to the fan with the flex connector placed on the larger side. This inlet modification allows air to enter the fan inlet from a larger diameter duct, reducing air turbulence and velocity.

Sudden Change in Duct Size

A Fan Coil Unit (FCU) mounted above the ceiling in an office space within a University research building generated noise levels reaching 70 dB when the fan was in operation. Not unpredictably, the office worker complained. An air balance report indicated that the FCU was operating far above the level for which it was designed (see Fig. 6.4a). In other words, the airflow, static and fan speed did not follow the classic "laws" of fan operations. Assuming that the airflow rates were

Fig. 6.3 Flex connector for air duct (courtesy of Ove Arup)

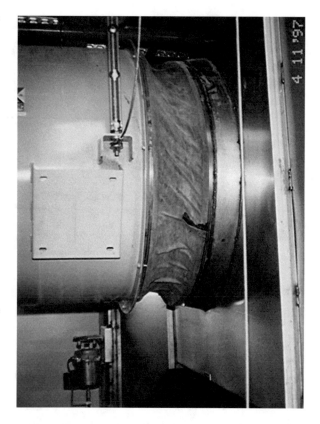

accurate, the fan speed should have been reduced to about 75 % of the original design value. The apparent discrepancy between measured and specified mechanical data (airflow, static pressure, and fan speed) suggested that the fan was operating with higher static pressure than actual measurements due to poor aerodynamic conditions, i.e., system effect. Site inspection revealed that the FCU was connected to a plenum duct causing a sudden drop in air velocity, by a factor of almost six times. The fan is a centrifugal forward-curve-blade unit with a small outlet opening, requiring gradual transition to a larger duct opening in order to meet the manufacturer's published performance expectations.

By installing a new transition duct at the fan discharge and adding a duct silencer between the FCU and the room air supply diffuser (see Fig. 6.4b) the expected performance levels were achieved. Post modification noise readings showed a significant reduction in the FCU noise levels of up to 10 dB.

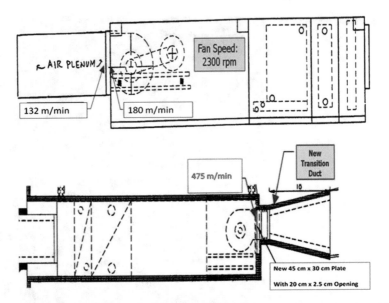

Fig. 6.4 AHU outlet design (*top*) before modifications and (*bottom*) after modifications (sketch drawing: courtesy of Ove Arup)

Large Induced Draft Fan Noise Impacts on the Neighboring Residential Community

A large micro-electronic fabrication facility with around-the-clock operation employed considerable mechanical equipment connected with the fabrication process. This included large scrubbers and exhaust fans and other equipment. The facility was located close to an existing residential community.

Scrubber exhaust fans are inherently noisy, primarily due to high static pressure on the fan operation. In most situations, scrubber exhaust fans are pressure blowers with high static pressure and relatively low flow-rate operation. Pressure blowers usually produce strong tonal sound that is centered on the blade passage frequency, 125–250 Hz. Site noise surveys taken close to the scrubber fans and at the nearby residential block showed a measurable contribution to the general ambient noise level from operating these scrubber exhaust fans.

Field airflow and static pressure measurements of the scrubber exhaust fans indicated that the noisy exhaust fans were operating on the unstable region of the fan curve (meaning basically that the static pressure was much higher than the value used for the design) while the airflow rate was much lower than the design value. Several noise mitigation options were considered: (1) modifying fans to operate close to their design point; (2) employing active noise cancellation techniques to attenuate the tonal noise problem, and (3) fitting the exhaust stack with a "reactive" duct silencer. Issues such as the impact on the facility's operation, constructability,

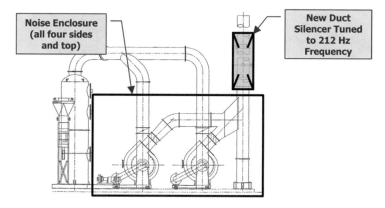

Fig. 6.5 Elevation inducted draft fans 1 and 2 (courtesy of Ove Arup)

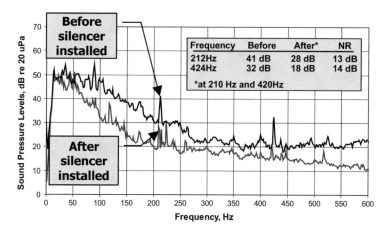

Fig. 6.6 Noise levels in neighborhood (courtesy of Ove Arup)

compatibility with the fan duct environment, and cost were considered for each of the mitigation measures. Additionally, noise mitigation had to allow for the least amount of downtime for the fan operation, be buildable and maintainable, and not restrict the fan operation.

Tuned sound attenuators (reactive duct silencers) were designed based on the measured frequency and amplitude of the noise problem. These silencers were tested at the factory for compliance with the project-specific requirements and were installed at the exhaust stacks (see Fig. 6.5).

Post-mitigation measurements indicate that the exhaust fan noise level in the neighboring community was inaudible and well-masked by the general ambient sound environment (see Fig. 6.6).

Acknowledgement Thanks to Amir Yazdanniyaz, PE with Acoustical Engineering Services, who integrated material from several sources into the initial draft.

Chapter 7
Electric Power Grounding and Conditioning

Abstract Minimization of electrical noise is a critical factor in the design and construction of all sensitive electrical instruments. This chapter addresses two important sources of electrical noise commonly found in laboratory equipment and hardware, and methods for their minimization. First, electrical noise generated from improper grounding of system components and associated hardware. Appropriate National Electrical Code is referenced for such terms as grounding conductors, grounded conductors, equipment grounding conductors and grounding electrode systems. Basic grounding requirements for every facility and a methodology for reducing electrical interference in facility design are provided. Second, common problems arising because of the instability of power source voltages, means to measure these power source variations and how they can be prevented by applying power conditioning devices to critical circuitry are described.

Grounding

Grounding is generally meant to be an electrical connection to earth. Hardware designers often connect circuits to earth to reduce the noise levels in their designs. This leads to the conclusion that good grounding is a way to limit electrical interference problems in facilities. For this reason, equipment designers often specify that special grounding arrangements be provided. Unfortunately, the word "grounding" has many meanings depending on the experience and background of the engineer using the word. Before we can discuss the role of grounding, we need a good working definition.

To illustrate the range of meanings, an analog engineer might refer to five or more different ground conductors in one circuit. These include input, output, power supply, equipment, chassis, and reference ground, to mention a few. These grounds may or may not be connected to each other or to earth.

To a power engineer, the word "grounding" refers to a connection to earth or its equivalent. In power wiring there are grounding conductors, grounded conductors,

© Springer International Publishing Switzerland 2015

A. Soueid et al. (eds.), *Buildings for Advanced Technology*, Science Policy Reports, DOI 10.1007/978-3-319-24892-9_7

equipment grounding conductors and grounding electrode systems. These word groups are carefully defined in the National Electrical Code[1] and imply specific conductors in the power wiring of a facility. These conductors must be present to meet the requirements of the Code.

Systems engineers involved in the interconnection of hardware worry about noise related problems. They often want a facility that follows their particular grounding philosophy. They may specify various reference conductors that should be used for connecting equipment grounds or signal shields. They assume that by using these grounding conductors they can limit or isolate interference. These conductors are often called clean, isolated, signal, analog, digital or military ground and are intended to carry any noise to a special "good" earth connection. Unfortunately this methodology does little to maintain a quiet facility. These invented terms are not defined in any electrical engineering book and they are not controlled by any regulatory agency. These invented expressions just add to the many possible meanings of the word "grounding." Providing these special conductors does not mean they will be beneficial. In fact, they can provide a false sense of security and little else, aside from raising the cost of construction.

A bit of history can help explain how some of these grounding ideas originated. In the 1950s, vacuum tubes were used and most electronic hardware was constructed in a single metal box. It was common practice to use single-point grounding inside the box. All the circuit grounds were brought back to a single point-grounding stud. It was also found that if this point were earthed to a water pipe the noise coupled into the circuit would usually be reduced. In an effort to limit expected noise problems many large facilities are actually wired using this single-point ground philosophy. A separate grounding well with a conductor buried in the earth that corresponds to the stud in a small item of hardware collects grounding conductors from the circuits in each building. While this grounding solution can work for a small piece of hardware, many problems arise when dimensions are large or facilities house many different electronic devices. This single point grounding practice should be avoided in today's high frequency environment; it is a move in the wrong direction.

As stated earlier, the general meaning of the word grounding is a connection to earth. In an aircraft, the frame of the aircraft can be called ground and it obviously is not connected to earth. Sensitive equipment can function in an aircraft by using the framework as the reference conductor. This observation leads to the question "Must the electronics in a facility be connected to earth?" In most cases the answer is yes. The reasons are that grounding provides for fire protection, shock avoidance and lightning protection. There are many ways to provide this grounding and still meet

---

[1] The National Electrical Code (NEC), or NFPA 70, is a United States standard for the safe installation of electrical wiring and equipment. It is part of the National Fire Codes series published by the National Fire Protection Association (NFPA) (source: http://en.wikipedia.org/wiki/National_Electrical_Code).

the requirements of the National Electrical Code. With the right approach, the facility power designer can use the Code so that there is minimum interference.

For purposes of this discussion, the definition for facility grounding is: the interconnection of conductors used in a facility together with the distribution of power that minimizes the generation of electromagnetic fields.

Basic Grounding Requirements for Every Facility

The National Electrical Code requires that one power conductor be earthed at the service entrance and used as the neutral conductor in three-phase power. This grounding provides an earth path for lightning that strikes outside of a facility. Inside a facility, many other conductors are earthed including building steel, buried gas lines and water lines.

To avoid any possibility of shock, every conductor carrying power must be insulated and/or mounted inside a metal housing. These housings must be connected together and earthed at the service entrance where the power is earthed. The conductors that ground the hardware are called the "green wires" or "safety wires" and are included with power conductors in all power runs. If a power conductor faults to a safety conductor or housing, a breaker must trip. The fault path must be low impedance (milliohms) to guarantee a large fault current. This is automatically the case if power is installed per the National Electrical Code.

In most facilities, power is associated with many of the large conducting fixtures or housings, such as racks of hardware, computer frames, boilers, motors, ducts, elevators, etc. Many of these conductors are earthed when they are installed. This system of conductors is called the grounding electrode system of the facility. Sections of a grounding electrode system should never be insulated and earthed separately. Connections between earth points are rarely below $10\,\Omega$ and this is too high a resistance for fault protection. If one of the earth connections were to fault to a "hot" conductor, a load current of 10 A would flow, not enough to trip a breaker. Plus, one ground would be 120 V from another nearby ground. This configuration creates a real shock hazard. For this reason, the practice is unsafe and illegal.

Stray Power Currents in Building Conductors

A utility company often feeds power for many users from the same transformer. The neutral conductor may be earthed at many points along the run, for lightning protection. Since most distribution power is three-phased, the load on each of the phases is not totally balanced. This means that some of the unbalance current flows to the neutral conductor. If the neutral conductor is earthed in multiple locations, then some of the neutral current flows to the earth. In areas with many users, there is apt to be significant neutral current that uses the earth as its conduit. Buried conductors provide a lower impedance path than earth, concentrating stray currents in areas of industrial activity. This problem can be avoided by using a separate distribution transformer for each facility.

Power Line Filters

Almost every piece of electronic hardware has a power line filter. These filters are connected between power conductors and hardware cabinetry, allowing some power line current and disturbances to flow to the grounding electrode system of the facility. Unfortunately, some of the current flows through building steel, racks, conduit, cable shields and safety conductors. A different grounding plan for the facility might redistribute the current, but it cannot stop filter currents from flowing. Power wiring that uses isolated equipment grounding conductors tends to limit the performance of the line filters. It also has the effect of forcing the current to find parallel paths through signal shields, creating interference. For this reason, isolated equipment grounding is not recommended.

A Few Common Misconceptions

Grounding Provides a Path for Return Current Flow So That It Does Not Flow in Sensitive Circuits

Fact: At typical AC frequencies, power current flows in loops. If current flows to earth it must find a path to return to the circuit. Earth does not sink the flow and cannot be treated like a noise disposal system. These undefined loops create an electromagnetic field that can couple into all other conductor loops in the area.

Unwanted Current That Flows Through Building Steel or Through a Conduit Is Not a Problem

Fact: This current is related to an electromagnetic field and a circuit loop. The field extends to all nearby conductors and implies current in these conductors. If these conductors carry signals, the result can be interference.

A Circuit of Ground Conductors Can Limit Interference in a Facility

Fact: All interference is coupled by fields. The fields from current flow can be limited if the conductor geometry is correct; but a circuit does not control or define the conductor geometry.

The Equipment Designer and the Facility

Most equipment designers understand that their hardware must function in a less than adequate environment, using power with nominal regulation and harmonic distortion. External sensing devices often define the quality of a measurement. If there is sensitivity to temperature, humidity or vibration, this can be a facility design issue. If there is sensitivity to the ambient electric or magnetic field then the problem might be a facility or equipment issue. Hardware, by its very nature, can generate electric and magnetic fields that can upset another piece of hardware in the same area. In a sense, this is a hardware compatibility issue. The facility can be designed with users in mind so that every user has a local electrical environment that is satisfactory. Some hardware may have to be physically isolated to limit radiation or to limit exposure to radiation. Some equipment may have to be in shielded rooms and others might require a special power feed.

Power Feeds: The Isolation Transformer

At frequencies below 100 kHz the electric field dominates when there are small currents flowing in the circuits. The electric field can be controlled by an enclosure made from a conductive material. A metal box connected to the circuit at one point can limit coupling to electric fields outside of the box. The box can be the size of a building, a room or a piece of hardware. Ideally, electric fields generated inside the box cannot get out and external fields cannot get in.

Power entry is the most common breach of this type of enclosure. Multiply shielded transformers can be used to provide some immunity from power related interference. One method of power isolation involves the use of a separate power source. This technique allows the secondary neutral to be re-grounded and avoids the coupling that often occurs because of shared power neutral connections. Providing separate power sources can be an important part of a facility design.

Magnetic Field Control

In installations involving power transformers or power wiring in close proximity, there can be a power related magnetic field. The shielding of magnetic fields at 60 Hz is a difficult problem, particularly when the field intensity must be kept very low. Shielding an entire area can be expensive as well as difficult. Shielding small volumes can be accomplished by enclosing them in layers of high permeability material and copper.

Methodology in Reducing Electrical Interference in Facility Design

1. Specifications: It is a good idea to request information regarding the sensitivity requirements of the various users. What physical separation is required? What electromagnetic environment is specified? Are there special power requirements? Should some users be physically isolated?
2. Neutral grounding: Power related earth current flow results in a low-intensity magnetic field. Distribution systems that limit neutral earth flow can control this problem. As an example, avoid supplying power from a single distribution transformer to several buildings.
3. Service entrance: Multiple sources of power should enter a facility at a single point. This limits neutral voltage drops in the grounding electrode system.
4. Utility grounding: If practical, the same entry point should be used for power, cable, telephones and auxiliary power.
5. Distribution transformers: It is best to mount transformers away from building steel. The leakage field around a transformer can cause currents to circulate in the entire building. The result is a magnetic field near every steel member.
6. Power wiring: All power wiring should be twisted where practical. Neutral conductors should be twisted with their associated power conductors to limit magnetic fields. It is never a good idea to allow one neutral to be shared between two circuits. Conduit is preferred to open trays for power distribution.
7. Distribution transformers: Distribution transformers that are near sensitive areas should be designed to have low-leakage inductance. This limits the load-related magnetic field that extends outside of the transformer.
8. Load separation: Loads for heavy equipment and other high current loads should be taken from separate transformers where practical. The feeder system should be designed to limit any load related interference.
9. Power routing: Conduit carrying power for heavy loads should not be routed near sensitive areas. Panels or switchgear that distribute this power should not be located on walls that enclose sensitive areas; distribution routing should not be left to chance.
10. Separately derived power: In applications involving sensitive electronics, a transformer can supply separately derived power, often called a power center. This transformer can be mounted on ground plane. For best results, it should be located near the associated loads, limiting many forms of interference.
11. Isolated grounds: The code permits separate equipment grounds for each receptacle; however, this practice should be avoided. Separate grounds for each receptacle limits the performance of hardware line filters and raises the general ambient noise level in a facility.
12. Lightning protection: Down conductors should not take sharp bends and multiple paths for lightning current should be provided. In desert areas, it is important that down conductors are extended into the earth. It is also important to protect against lightning hitting a high point such as a ventilation duct or an

antenna, and to provide building steel connections for ground plane installations. The grounding electrode system should connect to the lightning conductors at multiple points.

Note on Ground Planes

A ground plane can be used to limit fields coupling into cables that interconnect various electronic enclosures. This assumes that cable runs and the ground plane form small loop areas and it is only practical if the cables rest on the ground plane. Also, rack housings must be a part of the ground plane. This point is often poorly considered in systems layout.

It is bad practice to earth a ground plane at only one point. Ground planes should be connected to the grounding electrode system at many points. Grounding the electrode system at many points is the only way to effectively limit field coupling into hardware, and it is also the best way to provide lightning protection.

Voltage Stability

A March 30th, 1998 Internet article in *Semiconductor Business News* stated that interruptions in semiconductor manufacturer processes can cost as much as $2 million dollars revenue per day. Interruptions can be due to voltage sags caused by ice storms, floods, hurricanes, or lightning; power distribution equipment failures; or other system anomalies. Typically described in terms of magnitude and duration (see Fig. 7.1), voltage sags affect the operation of sensitive production equipment, leading to shutdown or malfunction, (and lost productivity and revenue). If this occurs during normal power system events, the equipment is incompatible with its electrical environment.

The SEMI F47 Voltage Sag Standard

SEMI F47[2] was developed to improve system compatibility for semiconductor tools (Fig. 7.2). For over 2 years, members of the Semiconductor Industry Electric Power Research Institute (EPRI) worked together with their industrial customers

---

[2] SEMI F47 is an industry standard for voltage sag immunity. It says that industrial equipment must tolerate voltage sags, or dips, on the AC mains supply to specific depths and durations. It is such a good and useful standard that many other industries use it, either formally or informally. http://www.powerstandards.com/semif47.htm

Fig. 7.1 Voltage sags are described by magnitude and duration (copyright International SEMATECH, Inc. 1999. Reprinted with permission from SEMATECH. All rights reserved)

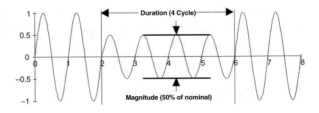

Fig. 7.2 SEMI F47: voltage sag ride-through capability curve (copyright International SEMATECH, Inc. 1999. Reprinted with permission from SEMATECH. All rights reserved)

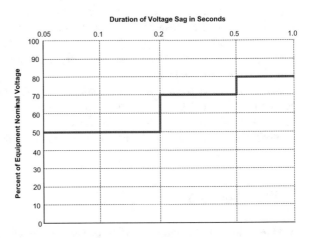

and equipment suppliers bring the standard to fruition in 1999.[3,4] The standard was issued in February 2000 and is in place today as version F47-706. In order to make semiconductor/nanotechnology equipment meet this standard, one can take two approaches. The first is to build in voltage sag immunity by use of DC powered control systems and proper design techniques. The second method is to use low cost power conditioning at the control voltage level.

Control-Level Approach

Control-level solutions involve identifying components or circuits in manufacturing process equipment that are sensitive to voltage sags, and protecting only those. When process equipment shuts down as a result of voltage sag, the weakness is usually found in a sensitive control component in the control circuit. Often, the ride-through of a semiconductor tool is directly related to the ability of one or more

---

[3] M. Stephens, "Semiconductor Equipment voltage Sag Immunity Improvements," EPRI PEAC Corporation. http://www.F47testing.com (2002).

[4] M. Stephens, D. Johns, J. Soward, and J. Ammenheuser, "Guide for the Design of Semiconductor Equipment to Meet Voltage Sag Immunity Standards," International SEMATECH Technology Transfer #99063760B-TR (1999).

small components to survive a voltage sag event. In this case, the use of selective power conditioning can lead to a great improvement in the overall tool's robustness and resistance to voltage sags.

EPRI research has shown that correctly protecting control components and circuits can increase the operating envelope of any equipment. It is only necessary to protect the single-phase-powered "weak links" in the tool, since all equipment power users are not ultra-sensitive to voltage sags and do not require conditioned power. The loads that are typically fed by selective power conditioning devices are single-phase devices with voltage requirements from 100 to 230 Vac. Figure 7.3 shows a control-level solution where the control circuit is protected with a small power conditioner. Additionally, the cost of a control-level solution is about 5–10 % of the cost of an equally effective panel-level solution.

Many voltage-sag related tool shutdowns can be prevented by applying power conditioning devices to critical circuitry. Critical items for most tools are the emergency off circuits, control power, critical instrumentation, DC power supplies and controller (computer) power. Often fed by single-phase voltage, there are several options for improving the ride-through in these areas. Typical selective power conditioning devices are shown in Fig. 7.4.

The Voltage Dip Proofing Inverter

The voltage dip proofing inverter (DPI) falls into a class of device referred to as battery-less ride-through devices (BRTD). Since the DPI operates only when the voltage sag is detected (off-line technology) it only needs to be sized for the nominal load. The device basically continually rectifies incoming AC voltage to charge the DC bus capacitors. When voltage sag is detected below an adjustable threshold, the line to the incoming power to the device opens and the DPI supplies a square-wave output to the load for about 1–3 s. The amount of time the load should be supplied can be calculated based on the real power and energy storage of the particular DPI.

Constant Voltage Transformer

The constant voltage transformer (CVT), or ferroresonant transformer, maintains two separate magnetic paths with limited coupling between them. Output contains a parallel resonant tank circuit and draws power from the primary to replace power delivered to the load. The transformer is designed so that the resonant path is in saturation while the other is not. As a result, a change in the primary voltage will not translate into changes in the saturated secondary voltage. CVTs allow for much better voltage sag ride-through if they are sized to at least two and a half the

Fig. 7.3 Control-level
power conditioning
approach (EPRI 2003)

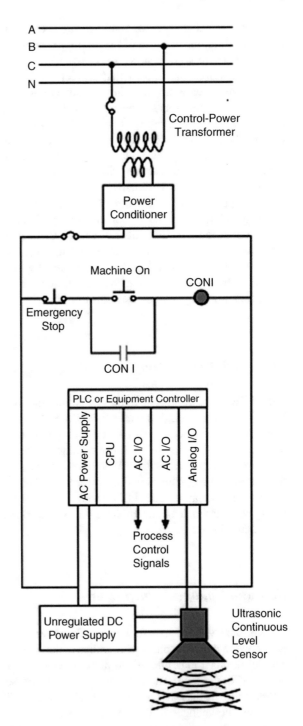

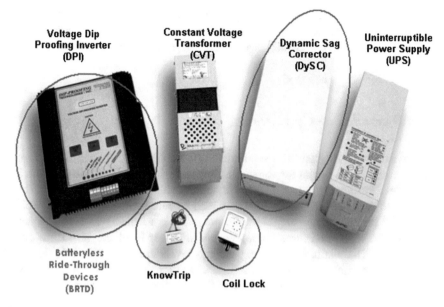

Fig. 7.4 Common selective power conditioning devices (copyright International SEMATECH, Inc. 1999. Reprinted with permission from SEMATECH. All rights reserved)

nominal VA requirement. Oversized in this manner, CVTs can supply 100 % of nominal voltage when the input voltage has dropped as low as 40 % of nominal.

Uninterruptible Power Supply

Uninterruptible power supply (UPS) comes in three basic types: standby, line-interactive, and rectifier/charger. Standby UPS switches to a battery and provides an inverter output to the load once voltage sag is detected. If the transfer is fast enough (less than one per cycle) and is in phase with the incoming voltage, typical control components are not likely to be affected by the sag event. It is important to carefully select this type of UPS in order to guarantee that sensitive control loads will not drop out before the unit switches to the inverter. Line-interactive UPS is an on-line device that employs a regulating transformer (CVT) when the incoming voltage is nominal. When voltage sag is sensed, the unit switches to the inverter to power the load. High in-rushing loads must be taken into account when using this unit since the CVT output can collapse from overloading. Rectifier/charger UPS is also an on on-line unit that constantly rectifies the incoming AC line voltage. The resulting DC voltage is then used to charge the batteries and to feed the inverter circuit for the unit's output section. In the event of voltage sag or outage, the unit switches to the battery for the source of the inverter's power.

Ultimately, the determination of whether to use a UPS or some other voltage conditioning device depends on whether the load requires power during a brief outage and the end user's willingness to perform periodic maintenance on the unit's batteries.

The Dynamic Sag Corrector

The dynamic sag corrector (DySC) system is a BRTD that corrects voltage sags; it supplies sine-wave output for voltages as low as 50 % of nominal, lasting from 3 to 12 cycles. By drawing power from the remaining voltage, the DySC injects a series voltage to regulate the output. The units can be fitted with capacitors as well to allow for limited outage ride-through, comparable to the DPI. This product is available in single and three phase designs in power levels ranging from 1.5 to 2000 kVA. The available operating voltage levels are 120, 208, 240, 277, and 480 Vac depending on the model used. The DySC was developed in tandem with the SEMI F47 standard and is targeted to the semiconductor industry.

Coil Hold-In Devices

Coil hold-in devices are also BRTD that are designed to prop up individual relay and contactor loads. Two brands found on the market are the KnowTrip[5] and the Coil Lock[6]. These units are designed to mitigate the effects of voltage sags on individual relays and contactors. Typically, the coil hold-in device is connected in line with the incoming control signal for the relay or contactor. Available for coil voltages of 120, 230, and 480 Vac, the best application for this device is to prop up relays and contactors that are in an EMO (Emergency Machine Off) master control relay, or motor control center circuits. Costing less than 50 dollars, these units are very economical to support contactors and relays. Typical coil hold-in devices allow a relay or contactor to remain engaged until the voltage drops to around 25 % of nominal. The unit is installed between the relay or contactor coil connection terminals and the incoming AC control line.

---

[5] Contactor dropout, because of power line brownouts, will be eliminated when a KnowTrip® is added to the contactor. The KnowTrip® is a patented device manufactured by SCR Controls, Inc., Matthews, NC, under license from Duke Energy. http://www.scrcontrols.com/products.asp?productName=KNOW%20TRIP&page=specs

[6] The Coil-Lock electronics works with AC relays, contactors, and solenoids as a hold-in-device that keeps motors and other critical process elements on line during those annoying momentary voltage sags. http://www.pqsi.com/coillocks.html

Acknowledgement Thanks to Ralph Morrison, Mark Stephens, and Chuck Thomas who synthesized material from several sources into the initial draft.

Bibliography

H. Akagi, E.H. Watanabe, M. Aredes, *Instantaneous Power Theory and Applications to Power Conditioning* (Wiley, Hoboken, NJ, 2007). ISBN 0470107618

R.C. Dugan, M.F. McGranaghan, S. Santoso, H.W. Beaty, *Electrical Power Systems Quality*, 2nd edn. (McGraw-Hill, New York, 2003). ISBN 0-07-138622-X

L.L. Grigsby (ed.), *Power Systems*, 2nd edn. (Taylor & Francis, Boca Raton, FL, 2007). ISBN 0849392888

R. Morrison, W.H. Lewis, *Grounding and Shielding in Facilities* (Wiley Interscience, New York, 1990). ISBN 0-471-83807-1

K.K. Sen, M.L. Sen, *An Introduction to FACTS Controllers: Theory, Modeling, and Applications* (Wiley, Hoboken, NJ, 2009) (FACTS—flexible AC transmission systems). ISBN 978-0-470-47875-2

R. Sittig, P. Roggwiller, *Semiconductor Devices for Power Conditioning* (Plenum, New York, 1982). ISBN 978-0306411311

The EPRI "Control-Level Power Conditioning Approach" illustration, taken from "Extending the Operating Envelope of Existing Manufacturing Processes: Control-Level Solutions", EPRI's "Tech Update 4-Tech Issue" (Feb. 2003), product ID: 1001664

A. von Meier, *Electric Power Systems: A Conceptual Introduction* (Wiley, Hoboken, NJ, 2006). ISBN 978-0471178590

Chapter 8
EMI/RFI: Electromagnetic and Radio-Frequency Interference

Abstract A myriad of man-made and extra-terrestrial sources in the non-ionizing electromagnetic (EM) energy range (0 Hz to 750 THz) emanate electromagnetic and radio-frequency interference (EMI/RFI). All sources with emissions throughout this broad frequency range can potentially degrade the performance of high-resolution imaging instruments such as SEM, TEM, FIB, and STM. This chapter discusses: (a) the most effective methods to measure ambient EMI/RFI emission levels around proposed and existing building sites; (b) units of measurement and susceptibility; (c) recommended minimal EMI/RFI thresholds for scientific tools; and (d) AC Extremely Low Frequency (ELF) magnetic flux density simulations at power frequencies. Finally, corrective strategies and costs to attenuate and control elevated EMI/RFI environments will be presented including active cancellation systems, zero milligauss shielding systems, Rigid Galvanized Sheet (RGS)/Electrical Metallic Tubing (EMT) conduits for electrical power distribution, and other mitigation techniques.

Introduction

A myriad of man-made and extra-terrestrial sources in the non-ionizing electromagnetic (EM) energy range (0 Hz to 750 THz) emanate electromagnetic and radio-frequency interference (EMI/RFI).[1] The interference comes from continuous and spurious emitters that compromise and degrade the performance of high resolution imaging systems (SEM, TEM, FIB)[2] and other equipment. These include diagnostic medical equipment (iEEG, EKG, EMG, MRI),[2] scientific instruments, and computer systems in nanotechnology, clinical and scientific research facilities. EMI/RFI that exceeds the manufacturer's susceptibility thresholds must be

---

[1] http://en.wikipedia.org/wiki/Electromagnetic_interference

[2] SEM scanning electron microscope; TEM transmission electron microscope; STEM scanning transmission electron microscope; FIB focused ion beam (writer); iEEG intracranial electroencephalography; EKG electrocardiography; EMG electromyography; MRI magnetic resonance imaging; NMR nuclear magnetic resonance (spectrometer).

© Springer International Publishing Switzerland 2015
A. Soueid et al. (eds.), *Buildings for Advanced Technology*, Science Policy Reports,
DOI 10.1007/978-3-319-24892-9_8

mitigated for optimal tool performance. In the U.S., standardized EMI/RFI susceptibility and interference testing is not mandated by law. Nor is it required by any of
the standards set by industry organizations such as the American National Standards Institute (ANSI) or the Institute of Electrical and Electronic Engineers
(IEEE). Susceptibility and interference measurements are needed to guarantee
compatible environmental conditions for highly sensitive scientific instruments
near EMI and RFI sources. The Federal Communications Commission (FCC),
Part 15, regulates RF interference from intentional and unintentional emitters.[3]
However, it does not address susceptibility issues directly, leaving much to be
desired when addressing existing building sites. Confusion abounds as each manufacturer presents their own method to measure and document the ambient
EMI/RFI environmental conditions required to ensure optimal tool performance.

This chapter will discuss: (a) the most effective methods to measure ambient
EMI/RFI emission levels around proposed and existing building sites; (b) units of
measurement and susceptibility; (c) recommended minimal EMI/RFI thresholds for
scientific tools; and (d) AC Extremely Low Frequency (ELF) magnetic flux density
simulations at power frequencies. Finally, corrective strategies and costs to attenuate and control elevated EMI/RFI environments will be presented including active
cancellation systems, zero milligauss shielding systems, Rigid Galvanized Sheet
(RGS)/Electrical Metallic Tubing (EMT) conduits[4] for electrical power distribution, and other mitigation techniques.

Electromagnetic Spectrum

Confusion runs rampant in discussions involving "what is" electromagnetic
(EM) and radiofrequency (RF) interference. The non-ionizing Electromagnetic
Spectrum shown in Fig. 8.1 ranges from DC 0 Hz to 750 Terahertz (THz).

Time-varying and transient magnetic fields (0.1 Hz to 10 MHz bandwidth) from
various sources—including moving subways, elevators, power sources and con-
verters, as well as high-power RF emission sources—generate electromagnetic
induction when the magnetic fields couple with conductive objects. Conductors,
such as electronic equipment, scientific instruments, and people, induce currents
and voltages within objects, causing "interference." In unshielded susceptible
electronic equipment and signal cables, electromagnetic induction generates electromagnetic interference (EMI). It is manifested as visible screen jitter in displays,
hum in analog telephone and audio equipment, lost sync in video equipment, and
data errors in magnetic media or digital signal cables.

The same occurs in research tools and diagnostic medical imaging systems that
are sensitive to (a) elevated EMI emissions, (b) emissions caused by extremely low

---

[3] Title 47 Code of Federal Regulations, Part 15, Radio Frequency Devices.

[4] http://en.wikipedia.org/wiki/Electrical_conduit

THE ELECTROMAGNETIC SPECTRUM

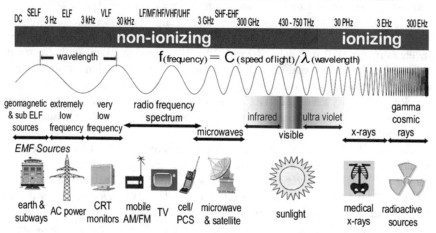

Fig. 8.1 Schematic of the electromagnetic spectrum (from VitaTech Engineering, LLC)

AC frequency (AC ELF) power, and (c) temporal perturbations to otherwise static magnetic field sources (DC EMI).

The bandwidth of radiofrequency interference (RFI) ranges from 9 KHz to 300 GHz. Far-field RFI sources typically radiate from electric field, planewave and microwave emitters. Typical problematic RFI sources include HF transmitters (e.g., amateur radio hobby users), mobile and marine transceivers, AM/FM radio antennas, TV and digital emitters, terrestrial microwave, cell phones, wireless networks and radar sources.

Recommended AC ELF & DC EMI & RFI Performance Requirements

Placement of each scientific tool and instrument depends on the actual EMI susceptibility under defined thresholds, which are often not easy to ascertain from the manufacturer's performance criteria. AC ELF and DC magnetic flux density susceptibility can be specified in magnetic field strength (amperes per meter—A/m) or one of three magnetic flux density terms: B_{rms}, $B_{peak-to-peak}$ (B_{p-p}) and B_{peak} (B_p) according to the conversion formula below:

$$B_{rms} = \frac{B_{p-p}}{2\sqrt{2}} = \frac{B_p}{\sqrt{2}}$$

To convert magnetic field strength to units of milligauss (mG), simply multiple the magnetic field strength by 4π. For example, 3 A/m is equal to 37.7 mG.

High-resolution imaging tools such as the JEOL 2010 transmission electron microscope require 0.2 mG horizontal and 0.3 mG vertical B_{p-p} (less than 0.18 B_{rms}) levels, so 0.1 mG and less in EMI Quiet Labs is a reasonable requirement. New super STEM (Scanning Transmission Electron Microscope) tools can demand levels as low as 0.01 mG B_{rms}. Using AC ELF magnetic flux density ambient data recorded at a site in question, along with simulated emission profiles, (from electrical conduits and switchgear), and the correct conversion formula, it is possible to identify the appropriate levels acceptable for each tool with one caveat: the correct EMI susceptibility figure must be ascertained from the manufacturer's specifications.

This is where the EMI challenge lies. Generally, for AC ELF sources, the minimum EMI threshold is 10 mG in unshielded electronic equipment, especially for 14″–17″ CRT color computer monitors and analog signal cables. However, the AC ELF EMI threshold for high-resolution 17″–21″ CRT color monitors is only 5 mG. Based on experience, the EMI AC & DC Magnetic Field Performance Specifications in Table 8.1 are recommended for NMR (Nuclear Magnetic Resonance) areas, cleanrooms, instrument and quiet labs:

The 0.1 mG B_{rms} for quiet and EM laboratories is recommended to accommodate the tool AC ELF and DC EMI performance requirements of current and next generation research tools. Achieving 0.1 mG B_{rms} levels without the use of AC ELF and DC magnetic shielding can be very challenging in most research facilities because of external and internal EMI sources. External sources include moving vehicles, loading dock activities, underground electrical feeders, etc. Internal EMI sources include elevators, electrical switchgear room, conduit runs, motors, and electric rooms and panels. Cleanrooms generally contain a significant quantity of environmental support equipment. This equipment generates high EMI emissions in close proximity to electromechanical equipment. Because of this, a slightly higher performance requirement of 0.3 mG B_{rms} is recommended; a practical number considering the complex EMI environments. There may be selected areas within the cleanroom where specific tools are located that require that the environment meet the 0.1 mG B_{rms} criteria.

Accurately predicting differences between 0.1 and 0.3 mG isolines from AC ELF and DC magnetic emission sources, considering the number of variables, is very challenging; especially when compared to measuring the actual EMI environment after a building is constructed and fitted with equipment and research tools. After a facility is operational, it is very difficult to measure the difference between 0.1 and 0.2 mG AC ELF and DC EMI levels due to the typical variations in electric power loads (caused by HVAC systems turning on and off, normal diurnal

Table 8.1 AC ELF & DC EMI Magnetic field performance specs

Ideal AC ELF & DC EMI Magnetic field performance specs	
NMR maximum requirement	1 mG B_{rms} (2.83 mG$_{p-p}$)
Quiet and EM laboratory maximum requirement	0.1 mG B_{rms} (0.3 mG$_{p-p}$)
Cleanroom maximum requirement	0.3 mG B_{rms} (0.85 mG$_{p-p}$)

variations in power demand, etc.) and nearby moving ferromagnetic masses (trucks of various sizes with variable loads, different elevator sizes and masses, etc.). Eventually, based on hands-on experience and tool use, the researchers will determine the optimal time of the day to perform complex and high-resolution experimentation with minimal EMI interference.

DC Electromagnetic Interference

Large and small ferromagnetic masses in motion such as elevators, cars, trucks, and metal doors produce quasi-static geomagnetic field perturbations in the sub-extremely low frequency (SELF) 0–3 Hz band. These perturbations radiate from the source, generating DC (EMI) in sensitive scientific tools and instruments. The magnitude of the geomagnetic field perturbation and radiated distance from the source depends on the size, mass and speed of the moving ferromagnetic object. Theoretically, DC magnetic emission sources (ferromagnetic objects, magnets, etc.) decay according to the inverse cube law. In practice, the decay rates are not ideal. Other problematic DC EMI sources include electromagnetic pulse (EMP) devices, which are usually high-voltage discharge instruments, subways, trolleys, NMRs, and MRIs.

Electron microscopes, Focused Ion Beam (FIB) writers and E-Beam Writers are also very susceptible to DC EMI emissions and require clean DC environments with 1 mG or less. Placement of scientific tools depends on the actual DC EMI susceptibility under defined thresholds, which are often not easy to ascertain from manufacturers' performance criteria. Electron microscopes are sensitive at 1 mG B_{rms} from DC disturbances, while SEMs and TEMs such as the JEOL 2010 have 0.4 mG horizontal and 0.2 mG vertical performance requirements. Next generation EM tools require less than 0.1 mG B_{rms} and Super STEMs (also known as ultra-high resolution STEMs) have a 0.01 mG DC EMI threshold. DC susceptibility in typical 1.5–4 Tesla MRIs can range from 1 mG to over 0.5 Gauss depending on the magnetic field strength, resolution, and type (open vs. closed, active shielding, etc.). Furthermore, to ensure a safe working environment around MRIs and NMRs, adequate signage must be posted at 5 and 10 Gauss lines to warn staff and visitors with implanted devices and to minimize inadvertent data corruption (coercivity) of credit cards and other valuable magnetic media.[5] Table 8.2 is a DC EMI threshold chart for EMI thresholds and magnetic media coercively.

---

[5] AC ELF magnetic field human exposure standards can be found at: NYS Public Service Commission as 200 mG @ 1 m on edge-row or 50 ft from 69 kV pulse; RPA/INIRC as 833 mG over 24 h, general public exposure; ACGIH 1000 mG, general public and workers with cardiac pacemakers; Swiss Bunderstat NCRP Draft Report as 10 mG form overhead/underground transmission/distribution lines, substations, etc.

Table 8.2 DC EMI threshold chart

DC EMI thresholds—CRT screen shift, noise and coercivity (data errors)	
<0.001 Gauss	SEM, TEM, e-Beam/FIB writers
0.75 Gauss	CRT Monitors and electronic instruments
5 Gauss	Cardiac pacemakers and implantable devices warning sign
10 Gauss	Credit cards and magnetic media warning sign
300 Gauss	Low coercivity mag-stripe cards
700 Gauss	High coercivity mag-stripe cards and video tapes

Recommended RFI Thresholds

Europe has developed susceptibility (radiated immunity) standards, such as the BS EN 61000-6-1[6]; in the U.S., ANSI has its C63 electromagnetic compatibility standards.[7] Engineers in the U.S. utilize the guideline: 3 V/m for the industrial RFI threshold and 1 V/m for the medical and scientific instrument RFI threshold. For maximum performance, these are reasonable requirements.

Defining Ambient EMI/RFI Emission Levels: The Site Survey

Every new building and laboratory project requires a site survey to document the existing levels of full-spectrum DC and AC ELF magnetic fields and wideband electric field RF ambient EMI/RFI. Quality, calibrated EMF instruments with minimum 0.04 mG AC ELF and 0.01 mG DC resolution are required to accurately record timed and contour/lateral path three-axis emission data around the proposed sites. Perimeter, lateral and contour three-axis AC ELF magnetic flux density data including the nearby roads, underground and overhead transmission and distribution lines, and any transformers and substations near the proposed site should be recorded. The AC ELF magnetic flux density data should be presented as Hatch, Profile and 3-D plots with timed data showing changes in demand and loads.

An interesting AC ELF magnetic flux density site survey was conducted at the Naval Research Laboratory. An underground CATV cable with ground and net current emissions was found to elevate emissions on site, as shown in Fig. 8.2a. The CATV cable was disconnected and replaced with fiber optic cable, showing an incredibly large reduction in magnetic field emissions to 0.05 mG on site, as shown in Fig. 8.2b.

---

[6] British Standards Institution, Electromagnetic Compatibility (EMC), BS EN 61000 http://shop.bsigroup.com/ProductDetail/?pid=000000000030211125; the International Standards Organization has electromagnetic compatibility standards for space (ISO 24637:2009) and health applications (ISO/TR 21730:2007).

[7] American National Standards Institute, Accredited Standards Committee C63, Electromagnetic Compatibility http://www.c63.org/index.htm

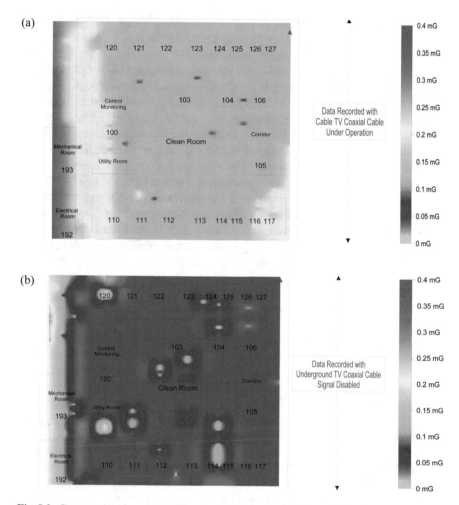

Fig. 8.2 Contour plot of resultant AC magnetic flux density levels recorded at 1 ft. intervals along a 10 ft. × 10 ft. grid on the site of the NanoScience Building at the Naval Research Laboratory. Data recorded with a calibrated FieldStar 1000 Gauss Meter. (**a**) With operating underground coaxial cable, (**b**) with the cable disabled (courtesy VitaTech Engineering)

Ground and Net Current Issues

Ground and net currents are due to violations of the National Electrical Code (N.E.C.)[8] in the electrical service, distribution and grounding systems of a building and violations of the National Electrical Safety Code (N.E.S.C.)[9] on distribution

---

[8] National Fire Protection Association, http://www.nfpa.org/aboutthecodes/AboutTheCodes.asp? DocNum=70

[9] IEEE Standards Association, http://standards.ieee.org/email/nesc2012.html

and transmission lines. Unbalanced phases on medium voltage distribution lines and 480 V/208 V low-voltage feeders generate zero-sequence currents, which return on the neutrals and grounding conductors. Most utilities maintain 5 % and less unbalanced phases on high-voltage transmission lines and 10–15 % unbalanced phases on distribution lines except in local neighborhoods where unbalanced phases may exceed 20 %. Some of the zero-sequence neutral currents on distribution lines travel along other electrically conductive paths, such as underground water pipe or earth channels, back to the substation. If all the zero-sequence currents were to return via the multi-ground neutral system (MGN) wire mounted on the pole under the three-phase conductors (sum of all phase and neutral currents are zero), then the magnetic fields would decay at the normal inverse square rate ($1/r^2$ in meters) from the single-circuit distribution line (same for transmission lines and low-voltage feeders). However, if only a fraction of the zero-sequence current returns on the MGN system or low-voltage neutral conductor, then there is net current missing (amount of current returning via other paths). This net current emanates a magnetic field similar to a ground current that decays at a linear $1/r$ (in meters) rate based upon the following formula:

$$B_{mG} = 2(I)/r \quad \text{where I is in amperes and r is in meters}$$

Magnetic fields from ground and net (zero-sequence) currents decay at a slow linear rate. The rate of decay of magnetic field strength versus distance from a current flow is illustrated in Fig. 8.3 using a 5 A ground and net current source. Distances from ground and net (zero sequence) currents are based upon a specified instrument performance criteria (i.e., 1 mG, or 0.1 mG).

Ground and net current magnetic field emissions are difficult to shield using flat or L-shaped ferromagnetic and conductive shields—the most effective shielding method for AC ELF ground and net current emissions requires a six-sided, seam-welded aluminum plate shielding system with a waveguide entrance. Low-ambient magnetic field levels can be achieved inside a research laboratory and imaging suite by adhering to the National Electrical Code and good wiring practices. However, these low levels can only be achieved under the most pristine conditions and without any circulating ground and net currents present on the primary electrical distribution system outside of the building, low-voltage 480/208 V distribution feeders and branch circuits inside the building systems, and the grounding system; otherwise, AC ELF magnetic shielding is required to obtain the performance objectives.

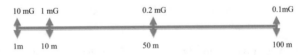

Fig. 8.3 Magnetic flux density level as a function of distance from a 5 A ground and net current source

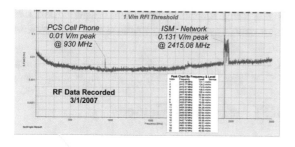

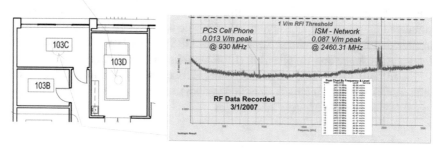

Fig. 8.4 RFI spectral data recorded inside a future TEM room (courtesy VitaTech Engineering)

Moving vehicles generate various types of DC EMI emission signatures and spikes based upon overall mass, speed and power source (e.g., electric trains emanate higher emissions compared to diesel trains). Although the ideal decay for DC sources is the inverse cube, the actual decay formulas for a bus based upon the recorded data was a magnetic field intensity in mG = 30 D-1.36, where D = distance from the source in meters.

It is absolutely essential to record the far-field electric field strength levels at selected locations around the proposed building site and inside labs in volts-per-meter (V/m) with a spectrum analyzer and calibrated antenna. An example recorded at Yale University's imaging laboratory site is shown in Fig. 8.4. Generally, more peaks appear from outside broadcast (FM & TV) and cellular (General System for Mobile Communications (GSM),[10] Personal Communications Service (PCS),[11] etc., emitters compared to the RF data (PCS and network) recorded inside a building.

Every project should include a final EMI/RFI site survey and report that records magnetic flux density data in 2-D and 3-D graphics overlaid on the site plan and floor plans, as well as a spectral analysis of the RF environment. The report should discuss the potential EMI/RFI impact of ambient data on the new facility and future research tools, and recommend a remediation strategy. These remediation strategies

---

[10] http://en.wikipedia.org/wiki/GSM

[11] http://en.wikipedia.org/wiki/Personal_Communications_Service

could include magnetic and RF shielding, increased separation distance from EMI sources, and active-cancellation technology. The report should also include associated cost estimates.

Schematic Design EMI/RFI Assessment Methods

During the Schematic Design phase, it is important to a meet with the client to discuss the EMI/RFI threshold criteria for laboratories, cleanrooms, and offices relative to the electronic and scientific instruments the spaces will house. A simulated magnetic flux density emission profile of a very large conduit ductbank in Shielded Rigid Galvanized Steel (RGS or GRS) in the Harvard Laboratory for Integrated Science and Engineering (LISE) building is shown in Fig. 8.5.

Based on the simulated emission profile, Harvard decided to use RGS conduits with a dual-substrate ductbank shielding system to ensure levels of 0.1 mG and less in the adjacent EMI-sensitive laboratories.

PVC, EMT, and RGS Conduit EMI Simulations and Comparisons

Simulating overhead and underground transmission and distribution lines under average and peak loads provides critically important separation distances between the various EMI sources and the potential EMI victims. These include EMI sensitive research tools, medical and diagnostic devices, occupants and visitors in laboratories, cleanrooms, hospitals, clinics, and nanotechnology facilities.

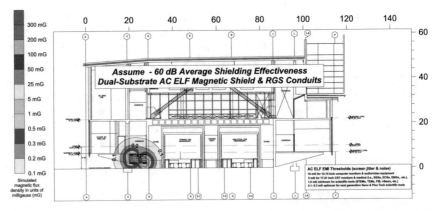

Fig. 8.5 Dual substrate shielded RGS conduit shield simulations (courtesy VitaTech Engineering)

Simulations are a cost-effective and accurate tool to assess the potential EMI impact from existing and planned electrical power sources located outside and within a building.

The shielding factor (SF) is the ratio between the unperturbed magnetic field B_o and the shielded magnetic field B_i as expressed in: $SF = B_i/B_o$ or in decibels $SF_{dB} = 20 \log_{10} (B_i/B_o)$. Normally, Rigid Galvanized Steel (RGS or GRS) conduit provides a minimum of -23 dB (0.07 SF multiplier) of shielding attenuation and Electrical Metallic Tubing (EMT) conduit provides -13 dB (0.22 SF multiplier). Polyvinyl chloride plastic (PVC) conduit and mineral-insulated (MI) cable have the shielding attenuation of 0 dB (1 SF multiplier), which is unity or none. All 13.8 kV primary feeders inside or within 40 ft. of the building must be in RGS conduits. All 480/277 V and 208/120 V high-current feeders inside and within 12 m of the building must be in RGS conduits. In circuits 100 A and less there is a marginal difference in the effective attenuation factor between RGS and EMT conduit. In circuits above 100 A, however, the wall thickness of the RGS provides a needed -10 dB of additional attenuation for high current circuits.

Magnetic field emission profiles from RGS and EMT conduits of different diameters and conduit wall thicknesses, conductor sizes, loads, and separation distances between conductors (twisted verses separated) have been simulated. Straight run single conduit simulations are easiest. Multiple conduits are more challenging because configuring the runs into the models takes so long. The minimum and maximum simulated emission depends on the angular location along a fixed radial distance from the conduit center under a balanced load.

Twisting the conductors and neutrals inside conduits improves the natural cancellation effect between three-phase and neutral currents. This cancellation effect is most pronounced in Polyvinyl Chloride plastic (PVC) conduit, especially at distances from 0.15 to 2 m from the conductors, as shown in Table 8.3.

Comparing the simulated twisted and separated (wide phase) emission profiles of 4-in. RGS conduit under 208 A balanced loads, the minimum and maximum levels converge at 2 m where there is virtually no difference between twisted and wide conductors (see Table 8.4).

Twisting the phase and neutral conductors decreases the emission profiles from PVC conduits and improves the performance of EMT conduits; however, the effect

Table 8.3 Comparison of magnetic field strengths from twisted and wide phase conductors in PVC Conduit (208 A balanced load)

PVC magnetic emission (twisted)			PVC magnetic emission (wide phase/separated)		
Distance (m)	Min (mG)	Max (mG)	Distance (m)	Min (mG)	Max (mG)
0.15	1054	1211	0.15	1478	2004
0.31	243	324	0.31	394	494
1	21	40.0	1	29	59
2	8.8	16.0	2	8.6	20
5	3.8	5.4	5	3.6	6.1
10	1.9	2.7	10	1.9	2.9

Table 8.4 Comparison of magnetic emission from twisted and wide phase conductors in RGS conduit

RGS Magnetic emission (twisted)			RGS magnetic emission (wide phase/separated)		
Distance (m)	Min (mG)	Max (mG)	Distance (m)	Min (mG)	Max (mG)
0.15	53	60	0.15	46	61
0.31	8.4	14.4	0.31	12.2	13.6
1	0.8	1.3	1	1.16	1.19
2	0.29	0.33	2	0.29	0.30
5	0.05	0.05	5	0.05	0.05
10	0.01	0.01	10	0.01	0.01

is minimal for RGS conduits. Twisting phase and neutral conductors in PVC and EMT conduits is recommended for all circuits. RGS for 100 A and higher circuits is recommended where the thick RGS steel wall improves the shielding effect, compared to the thinner wall EMT. Regardless, RGS should be used in all EMI-sensitive tool areas where magnetic field levels must be less than 0.5 mG.

Switchgear Simulation and Shielding

VitaTech simulated the 4000 A dual switchgear under 75 % loads in the proposed facility Center for Functional Nanomaterials (CNF) at Brookhaven National Laboratory. Based on the simulated data, the recommendation was to enclose the entire switchgear room with a dual-substrate AC ELF magnetic shielding system with −48 dB of attenuation. This reduced the magnetic field emissions on the nearest laboratory walls to 0.1 mG (see Fig. 8.6). The final performance EMI site survey verified the AC ELF magnetic flux density levels were 0.05 mG and less within the research laboratories adjacent to the main switchgear room. Shielding main switchgear rooms is essential when laboratories with EMI-sensitive instruments are located near high magnetic field emission sources.

The simulated AC ELF magnetic flux density emission profile from 400 and 800 A busways at the new NIST Metrology Facility in Gaithersburg, MD, operating at 100 % load and balanced phases is shown in Fig. 8.7. The simulation shows the necessity for carefully locating and shielding busways.

Elevator DC EMI Issues and Shielding

Passenger and freight building elevators generate serious EMI issues for sensitive imaging tools in nanotechnology and medical/diagnostic research instruments

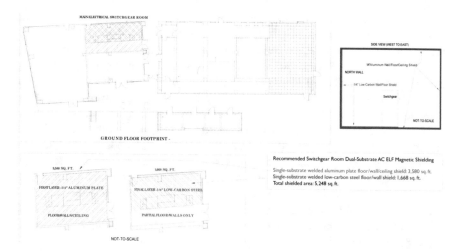

Fig. 8.6 Illustrative AC ELF magnetic shielding design for a main electrical switchgear room with dual-substrate shielding (two layers; note the different materials and locations as depicted in *red* and *blue*) (courtesy VitaTech Engineering)

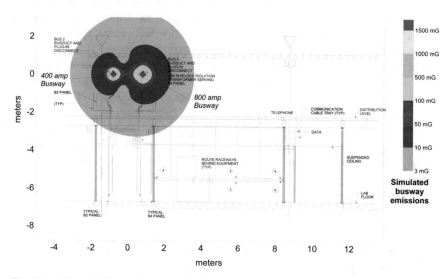

Fig. 8.7 400 and 800 A dual busway simulation of AC ELF magnetic emission @ 100 % rated load (courtesy VitaTech Engineering)

when located too close to the moving ferromagnetic mass.[12] Figure 8.8 shows the resultant (B_r) and component (B_x, B_y and B_z) rms recorded data at 10-m from a passenger elevator at the Birck Nanotechnology Center at Purdue University.

At 10 m (see Fig. 8.8) the passenger elevator maximum differential (dB/dt) $_{rms}$ changes are: 0.35 mG B_r, 0.34 mG B_x, 0.11 mG B_y and 0.33 mG B_z. Remarkably, a dB/dt differential DC EMI emission of only 0.1 mG B_r $_{rms}$ can emanate up to 27.5–30 m (90–98.4 ft.) from a passenger elevator depending on the elevator size and building construction type. VitaTech designed and installed the first DC EMI elevator shielding system in the world in the building AZBiodesign #1, Arizona Biodesign Institute, Arizona State University.[13] Elevator shielding systems are expensive ($275,000–$500,000) depending on the number of floors it serves, the shaft size, types of shielding and the number of applied layers of shielding required to achieve the final attenuation criteria. Figure 8.9 shows the actual recorded B_{rms} data at selected distances from the shielded passenger elevator in the AZBiodesign Phase #1 building.

The elevator DC magnetic shield provided between −10 and −15 dB of attenuation based on the distance from the center shaft and the proximity of other ferromagnetic materials to the fluxgate probe when recording timed DC emission data as the elevator moved in the shaft. Figure 8.9 includes four diagrams depicting the vertical motion of the elevator as a function of the sampled data with the average dB/dt differential change noted. A nearby MRI room measured only 0.1 mG dB/dt and less DC EMI levels at the center of the room. This is very low and acceptable for TEMs, SEMs, and all MRIs. Without the elevator shield the predicted levels would have ranged from 0.5 mG to over 1 mG dB/dt. The DC elevator shield provides adequate DC EMI protection for the MRI units and future scientific tools and instruments located inside the lower level and first floor laboratory areas. Shielding elevators is very challenging and a costly investment that should only be considered when there is minimal separation distance (less than 10 m) between the elevators and EMI sensitive research tools. The most cost-effective solution is to lock-out the elevators during high-resolution imaging research when using DC EMI sensitive tools located too close to the elevators.

Predicted NMR DC EMI Emission Profiles, Magnetic Shielding and Active-Field Cancellation

When an NMR magnet is energized, the ramp-up process emanates a differential magnetic field gradient as the unit achieves normal operating mode. This ramp sequence perturbs the geomagnetic field surrounding the NMR unit, quiet labs and

---

[12] The is a 10 mG EMI threshold for 12–15 in. color CRT monitors, computers and audio-video equipment; 5 mg EMI threshold for 17–21 in. color CRT monitors and sensitive instruments.

[13] http://www.biodesign.asu.edu

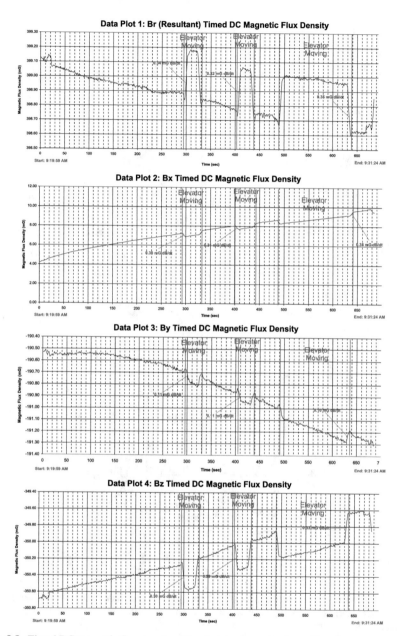

Fig. 8.8 Timed DC magnetic flux density at 1 m height above the floor, 10 m from the elevator, recorded at 1 s intervals, in the Birck Nanotechnology Center, Purdue University (courtesy VitaTech Engineering)

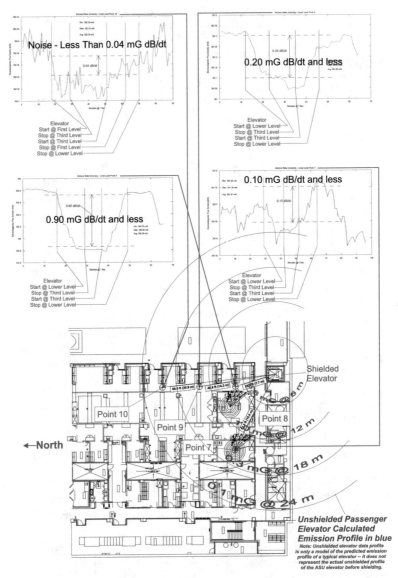

Fig. 8.9 Magnetic flux density at 1-m height from the Arizona State AZBioDesign lower level shielded elevator; timed DC $B_{r\ rms}$ taken at 0.5 s intervals with a MEGA FVM Vector Magnetometer (courtesy VitaTech Engineering)

cleanrooms. Once the normal operational condition is achieved, the magnetic field stabilizes and remains fixed around the NMR until deactivated, or powered down. The ramp-up and down sequence generates a changing dB/dt magnetic field that emanates into the surrounding areas, affecting EMI-sensitive instruments and tools.

It should be noted that this DC ramp gradient also radiates in the vertical direction at a similar inverse cube decay rate. There is a three-dimensional DC EMI gradient emanating from each NMR unit during the ramp-up procedure over 30 min until each unit reaches the normal steady-state condition, after which there are no changes in the DC magnetic fields.

Controlling the short-term EMI impact from NMR ramp sequence emissions is not simple and very expensive. Magnetic shielding can be effective in mitigating EMI problems during the ramp sequence, but is not justified by the expense (ranging from $150 K to $400 K depending on shield size) since the ramp sequence is only 30 min. The active Field Cancellation technology ($30 K to $60 K per room) that is normally installed around electron microscopes, and other EMI sensitive instruments, is not very effective in mitigating the EMI impact from the NMR ramp sequence. Each mitigation method has technical limitations and associated costs as discussed below:

- Magnetic shielding an NMR room (four walls, floor and ceiling; not the door) to control the differential dB/dt magnetic emissions from an NMR unit requires a significant amount of ferromagnetic material. Since cost is a factor, the minimal DC shield would require at least two different permeable ferromagnetic shielding materials composed of low-carbon steel plates and multiple layers of silicon-iron steel sheets in order to provide a shielding effectiveness ranging from 25 % (−2.5 dB) to 50 % (−6 dB). Higher shielding performances 75 % (−12 dB) to 90 % (−20 dB) are achievable, but the cost is not proportional, ranging from two to four times the cost of the −2.5 dB to −6 dB shield. This is why higher performing DC shielding for the NMR during the short term ramp sequence is not often recommended.
- Active field cancellation technology for the EMI DC-sensitive instruments located in quiet labs adjacent to the unshielded NMR units will not work. This is because the gradient levels on the shared wall with the NMR units will exceed the cancellation performance characteristics of most cancellation systems. Normally, active field cancellation technology is most effective responding to uniform DC and AC ELF sources from one direction, with a magnitude under a set stability threshold specified by the manufacturer. Transient DC sources, such as moving vehicles, elevators and ramp-up NMR DC fields, present a challenge because of the limited response time of the Helmholtz cancellation coils. Active cancellation systems can oscillate when exposed to multiple DC and AC ELF emission sources that exceed stability threshold parameters. This oscillation generates higher magnetic field levels inside the six active cancellation orthogonal Helmholtz coils used to generate the cancellation fields; as the active feedback system attempts to damper as the system oscillates. Another limit to active field cancellation technology is the use of a single three-axis sense probe, usually a fluxgate probe located by the EM column, which only responds to the highest magnetic field component measured at the probe. If there are several DC and AC ELF EMI sources emanating into the room protected by the cancellation system, the actual attenuation is limited to the bandwidth of the

system. Typically, active field cancellation systems can cancel the incident magnetic fields down to 0.5 mG per axis depending on the harmonic components and other complex characteristics of the incident magnetic fields. Nevertheless, active field cancellation technology is effective when magnetic shielding is not practical or cost effective.

When two or more NMRs are located in close proximity to one another, they must not intersect at the 5 Gauss DC isoline. MRI units must only intersect at the 1 Gauss DC isolines, otherwise imaging and interference problems will compromise optimal performance.

Near Zero-Milligauss AC ELF Magnetic Shielding Systems for EMI Sensitive Tools

At the Nanotechnology Center at the University of Arkansas at Little Rock, designers sought to design, install, test, and certify an AC ELF magnetic shield with a 0.1 mG $B_{r\ rms}$ final performance in a future SEM and TEM Electron Microscope room. An AC ELF EMF site survey of the EMI affected areas was conducted before shielding was installed. Figure 8.10a shows that the recorded B_r resultant magnetic flux density levels ranged from a mean average of 0.9 mG to a peak of 1.6 mG above the underground 13.8 kV distribution line with ground/net current emissions.

Floor-level data ranged from a mean average of 1.9 mG to a peak of 2.6 mG above the underground distribution line. Ground and net current is present on the 13.8 kV distribution line because the peak floor level is nearly twice the 1-m peak. The elevated magnetic field levels would generate EMI in both the future SEM and TEM Electron Microscopes (JOEL 2100 and JEOL 7000F 1K). Figure 8.10b shows the final flux density after installation of the shielding describe in the next paragraph.

The final shielding design depends on the following critical factors: maximum predicted worst-case 60-Hz magnetic field intensity (magnitude and polarization) and the earth's geomagnetic (DC static) field at that location; shield geometry and volumetric area; type and thickness of materials and their properties of conductivity (σ), permeability (μ), induction and saturations; number of shield layers; and, the spacing between sheet materials and layers. A Zero-Milligauss AC ELF EMF shielding dual substrate system was designed and installed on the floor, four walls and ceiling of the laboratory, which was 4.57 m by 6.1 m and 3.5 m high (15 ft. by 20 ft. and 11.5 ft. high). Figure 8.11 shows the actual seam welded high performance aluminum plate Zero Milligauss shielding system. The improvement is illustrated in Table 8.5.

Based on the recorded before (2.6 mG) and after (0.01 mG worst-case) AC ELF magnetic flux density data shown in Fig. 8.10, the calculated final Shielding Factor (SF) in decibels is as follows:

(a)

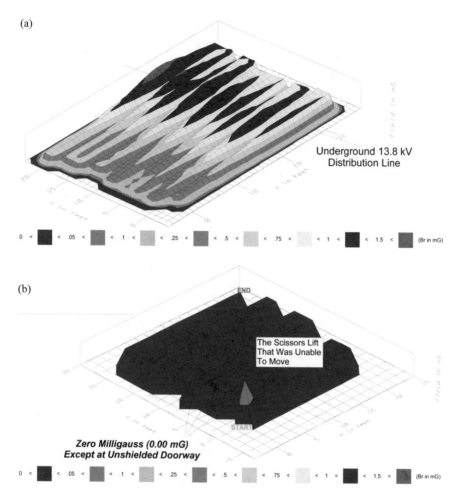

Fig. 8.10 Magnetic flux density levels at 1 m height in the microscopy room at the Nanotechnology Center, University of Arkansas, Little Rock, (**a**) before shield contour (lights off), and (**b**) after the shield was installed (courtesy VitaTech Engineering)

Fig. 8.11 Pictures of University of Arkansas Nanotechnology Center shielded spaces (courtesy VitaTech Engineering)

Table 8.5 Before and after shield magnetic flux density in electron microscope room

	Mean (mG)	Peak (mG)	EMI source(s)
Before (BS) contour full room at 1 m	0.89	1.6	
After (AS) contour SEM/TEM at 1 m	0.0004	0.04	Unshielded door

$$\text{SF}_{dB} = 20 \log \left(0.01\,\text{mG} / 2.6\,\text{mG} \right) = 20 \log \left(0.004 \right) = -48.3\,\text{dB}$$

A final Shielding Factor (SF) of -48.3 dB is very impressive. It is usually close to impossible to magnetically shield ground and net current sources, such as the underground medium voltage 13.8 kV distribution line beneath the future SEM/TEM room at this specific site, with 1–2 A of ground and net currents under average loads, and twice that during peak summer loads.

Near Zero-Milligauss AC ELF Magnetic Shielding System Options

Only a well designed and constructed passive AC ELF magnetic shielding system can attenuate ground and net currents, three-phase and higher frequency harmonic magnetic emission sources to 0.1 mG rms and below (actually to 0.01 mG) in the component B_x, B_y and B_z axis and the calculated B_r resultant. Nevertheless, it is possible to design and install a wide-spectrum (30 Hz to 30 GHz) AC ELF/RF Zero-Milligauss Shielding System with MRI RF doors, honeycomb RF filters, RFI power and signal filters for scientists who require very demanding laboratory environments. And for those sites with serious DC EMI problems from nearby elevators, vehicles, and other annoying quasi-static DC EMI sources, there is a high performance DC EMI shielding system that requires Active Compensation System (ACS) technology[14] coupled with DC shielding and the Zero-Milligauss AC ELF and/or RF Shielding System for the ultimate full-spectrum shield (0 Hz to 30 GHz) with ultra-low level shielding performance.

Final Full-Spectrum EMI/RFI Commissioning

All new nanotechnology facilities, hospitals, clinics, and research laboratories require a full-spectrum commissioning EMI/RFI site survey to verify compliance with the stated EMI/RFI performance requirements. Electrical power sources

---

[14] See, for instance, "Magnetic active compensation (MACS) for electron microscopy," J. Kellogg, Interference Technology, pages 1–4 (2003), www.ets-lindgren.com/pdf/kelloggi_03.pdf

should be near normal operating loads (or under load banks) to identify any ground and net current problems due to N.E.C. violations in the electrical distribution system. This is easily assessed with an AC ELF magnetic flux density site survey and survey wheel.

Commissioning should include recording of the perimeter, contour, and lateral AC ELF magnetic field data. Such recordings will reveal ground and net current emissions in electrical panels and conduits. If the elevated magnetic field bloom punches out from the low level ambient data along the survey path, electricians should look for the grounded neutrals and wiring errors to correct the problem, then resurvey to verify the problem has been corrected.

Record timed (at least 30 min per lab at 1 s intervals) AC ELF and DC magnetic flux density data in EMI-sensitive laboratories. Measurements should be synchronized with the computer clock (or data recording clock), then data should be recorded at the exact time the elevator moves from the lowest to highest floors, and back down several times during each recording.

Also, it is important to record DC EMI data in the EMI-sensitive laboratories from moving vehicles on the nearby streets, parking lots and other access areas. To accomplish this, a spotter, someone with a cell phone and synchronized watch standing outside of the building, calls in each moving vehicle.

It is also important to record contour, perimeter, and lateral magnetic flux density data within each EMI-sensitive laboratory and adjacent areas to identify any N.E.C. violations (ground and net current sources) and other EMI emission sources that will require remediation. Finally, wideband RFI data must be recorded with a calibrated spectrum analyzer to identify the various intentional and unintentional (spurious) RF emission sources inside and outside the building, including any special EMI/RFI-sensitive research laboratories. A detailed final commissioning report with graphics showing all recorded EMI/RFI data should be delivered to the client to document existing EMI/RFI conditions before the building and laboratory is occupied and operational.

Acknowledgement Thanks to Lou Vitale who synthesized material from several sources into the initial draft.

Bibliography

J. Carr, *The Technician's EMI Handbook: Clues and Solutions* (Butterworth-Heinemann, Boston, MA, 2000). ISBN 0-7506-7233-1

L.T. Gnecco, *Design of Shielded Enclosures: Cost-Effective Methods to Prevent EMI* (Butterworth-Heinemann, Boston, MA, 2000). ISBN 0-7506-7270-6

L.H. Hemming, *Electromagnetic Anechoic Chambers: A Fundamental Design and Specification Guide* (Wiley, New York, 2002). ISBN 0-471-20810-8

K. Kaiser, *Electromagnetic Compatibility Handbook* (CRC, Boca Raton, FL, 2005). ISBN 0-8493-2087-9

M. Mardiguian, *EMI Troubleshooting Techniques* (McGraw Hill, New York, 2000). ISBN 0-07-134418-7

H. Ott, *Electromagnetic Compatibility Engineering* (Wiley, New York, 2009). ISBN 978-0-470-18930-6

The ARRL RFI Book, 3rd edn. (American Radio Relay League, 2010), http://www.arrl.org/shop/The-ARRL-RFI-Book-3rd-Edition

D.R.J. White, M.F. Violette, *Electromagnetic Compatibility Handbook* (Van Nostrand Reinhold, New York, 1987). ISBN 9780442289034

Chapter 9
Airborne Contamination

Abstract This chapter provides a broad overview of the sources of airborne contamination that may occur in buildings designed for advanced technology instruments and fabrication facilities. International standards for classifying levels of cleanliness of the air in these spaces in terms of particle size and particle concentration are briefly described. This is followed by an overview of cleanroom design principles and the filtering systems employed to achieve the various standards of air cleanliness in the cleanrooms.

Introduction

Airborne contamination generally consists of:

- Particulates (organic and inorganic)
- Bacteria
- Metallic ions
- Airborne molecular contamination (AMC) consisting of vapors and gases.

Using electronics fabrication as an illustration, contamination has three primary effects on production. Killer particles can ruin individual chips, reducing product yield. Contamination can cause uncontrolled doping and premature electronic failure of chips. AMC can cause hazing of the lithographic optics, which is detrimental to the performance of the equipment.

There are many sources of contamination—both external and internal (see Fig. 9.1). There is the make-up air stream that is bringing in outside air. Depending on location, outside air can carry all sorts of contamination, including desert dust, sea spray, diesel fumes, etc. Production materials used in a fabrication facility, including gases, water and chemicals, can cause contamination. The wafers themselves and the carriers they come in are also sources of contamination, as are people working in the lab and their clothing. Not to mention construction, maintenance and cleaning people and materials. This is a lot to track and manage.

© Springer International Publishing Switzerland 2015
A. Soueid et al. (eds.), *Buildings for Advanced Technology*, Science Policy Reports,
DOI 10.1007/978-3-319-24892-9_9

Fig. 9.1 Sources of
contamination (courtesy of
HDR Architecture, Inc.)

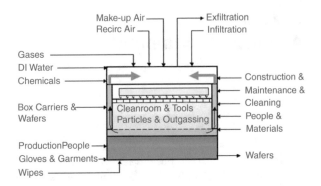

Table 9.1 Federal standard
209B—1976

Class name	Class limits (particles/ft.^{3})	
	>0.5 μm	5 μm
100	100	
10,000	10,000	65
100,000	100,000	700

The goal of manufacturing in a cleanroom is to minimize defects as well as other effects of contamination. The cleanroom should be a manufacturing environment that is clean enough without paying for cleanliness that isn't needed. This is why classifying levels of cleanliness is necessary.

The Federal Standard 209B document, circa 1976, provided a system for classifying three levels of cleanroom: Class 100,000, Class 10,000, and Class 100 (see Table 9.1). At that time, particle monitoring equipment was not sophisticated enough to detect lower levels.

By the time the Federal Standard reached revision E, several changes had taken place. First, semiconductor technology had advanced sufficiently so that Class 100 was not clean enough for high-yield manufacturing. Class 10 and Class 1 were added. Class 1000 was also added to round things out.

Additionally, 0.5 μm particles were larger than the critical particle size. Particle count limits were added for 0.3 μm, 0.2 μm, and 0.1 μm. Again, this was in response to the requirements of advancing technology.

Finally, the classification system went metric. Particle count limits were converted to particles per cubic meter, and a host of new metric classifications were added. All of these changes were the result of technology demands and globalization of the industry. The original three categories now comprise a very small part of the total system.

In November of 2001, Fed. Standard 209E was officially replaced by the international standard ISO 14644 which now has nine parts[1,2]:

---

[1] http://www.iso.org/iso/home.htm
[2] http://www.iest.org/i4a/pages/index.cfm?pageid=3322

- 14644-1: Classification of air cleanliness by particle concentration
- 14644-2: Specifications for monitoring and periodic testing and monitoring to prove continued compliance with ISO 14644-1[3]
- 14644-3: Test methods
- 14644-4: Design, construction, and start-up
- 14644-5: Operations
- 14644-6: Vocabulary
- 14644-7: Separative devices (clean air hoods, gloveboxes, isolators,[4] and minienvironments
- 14644-8: Classification of airborne molecular contamination
- 14644-9: Classification of surface particle cleanliness by particle concentration

ISO 14644 Parts 1 and 2 were revised in 2010 as draft international standards (DIS), see footnotes 1 and 2. The ISO system (see Table 9.2) adds two classes cleaner than Class 1 and, curiously, adds a class dirtier than Class 100,000. The old Fed Standard 209 Classes are shown for reference only; they are not a part of the new standard.

Cleanroom Design

Today, specifying ultra-high-efficiency, non-outgassing polytetrafluoroethylene (PTFE) filters is necessary. Filter systems are available with just about as many "9's" as you care to pay for: 99.9995 % or better. All materials in the cleanroom must be verified as non-outgassing. Special filtration systems are used to combat AMC, with continued use of old standbys such as positive pressurization and cleanroom certification, as well as cleanroom protocol—construction, operations and gowning. Contamination control efforts of cleanroom designers have proven effective. Two charts compare the sources of contamination between a 10-year-old Class 10 fab and a modern Class 1 fab (Fig. 9.2). Contributions to contamination from the cleanroom environment and from people have been significantly reduced.

Traditionally, there are a couple of general cleanroom designs commonly used. Figure 9.3 represents a ballroom arrangement where a production tool sits on a raised floor and is surrounded by a cleanroom environment. Recirculation air-handling units supply conditioned air to a supply plenum and then to the cleanroom filter ceiling system. The ceiling is composed of HEPA[5] or ULPA[6] filters. Air is returned to the recirculation units and make-up air is added to replace exhaust and

---

[3] http://en.wikipedia.org/wiki/ISO_14644-1

[4] http://en.wikipedia.org/wiki/Barrier_isolator

[5] High Efficiency Particulate Air filter, http://en.wikipedia.org/wiki/HEPA

[6] Ultra Low Particulate Air filter; http://en.wikipedia.org/wiki/ULPA

Table 9.2 ISO Standard 14644

ISO Classification	Class limits (particles/m³)						Fed 209 class
	0.1 µm	0.2 µm	0.3 µm	0.5 µm	1 µm	5 µm	
ISO Class 1	**10**	**2**					(0.01)
ISO Class 2	**100**	**24**	**10**	**4**			(0.1)
ISO Class 3	1000	237	1020	35	**8**		1
ISO Class 4	10,000	2370	10,200	352	**83**		10
ISO Class 5	100,000	23,700	102,000	3520	**832**	29	100
ISO Class 6	1,000,000	237,000		35,200	**8320**		1000
ISO Class 7				352,000	**83,200**	2930	10,000
ISO Class 8				3,520,000	**832,000**	29,300	100,000
ISO Class 9				**35,200,000**	**8,320,000**	**293,000**	

Fig. 9.2 Comparisons of demands of Class 10 and Class 1 cleanrooms (courtesy of HDR Architecture, Inc.)

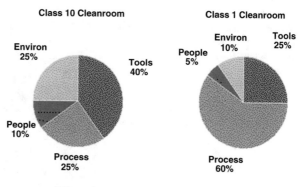

Fig. 9.3 Cleanroom with ballroom arrangement (courtesy of HDR Architecture, Inc.)

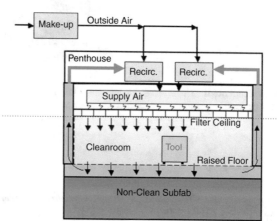

Ballroom With Non-Clean Subfab

provide positive pressurization. There are many variations on the theme, but the basic concept is standard.

The other traditional cleanroom arrangement is a bay and chase (Fig. 9.4).

In the bay and chase arrangement, cleanroom air is only introduced into the clean bays, which comprise a small percentage of the total fab area. Air is returned upward through the chase. The production tool is bulkheaded through the clean bay wall. This arrangement significantly reduces the airflow required to maintain cleanliness and therefore is used whenever the tool configuration allows.

Even with this configuration, a typical 300 mm fab will have over 60,000 m^3/min of recirculated air. Wafers are transported through the bay only and are exposed only to the clean environment. This is, of course, the purpose of the cleanroom—to provide a clean environment that reduces the wafer's exposure to contamination.

Fig. 9.4 Cleanroom with bay and chase arrangement (courtesy of HDR Architecture, Inc.)

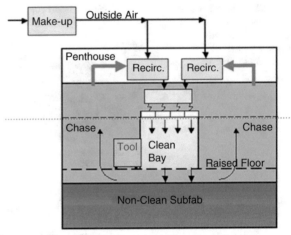

Bay and Chase With Non-Clean Subfab

Acknowledgement Thanks to Mark Jamison who synthesized material from several sources into the initial draft.

Bibliography

High Performance Cleanrooms, A Design Guidelines Sourcebook (Pacific Gas and Electric Company, CTM-0311-0875, January 2011)

C.K. Moorthy, *Contamination Control and Cleanrooms* (Anshan, Aldeburgh, 2004). ISBN 1-904798-06-3

C. Muller, Chapter 41: Airborne molecular contamination, in *Semiconductor Manufacturing Handbook*, ed. by H. Geng (McGraw Hill, New York, 2005). ISBN 0-07-144559-5

R.V. Pavlotsky, S.C. Beck, Chapter 38: Cleanroom design and construction, in *Semiconductor Manufacturing Handbook*, ed. by H. Geng (McGraw Hill, New York, 2005). ISBN 0-07-144559-5

M. Ramstorp, *Introduction to Contamination Control and Cleanroom Technology* (Wiley, Weinham, 2000). ISBN 3-527-30142-9

W. Whyte, *Cleanroom Technology: Fundamentals of Design, Testing and Operation* (Wiley, Chichester, 2001). ISBN 0-471-86842-6

Chapter 10
Bio-containment

Abstract This chapter compares and contrasts the design of nanotechnology cleanrooms and bio-containment Biological Safety Laboratories. As examples, suiting up in a cleanroom is necessary to protect product or test specimens, while suiting up in a bio-containment facility is necessary to protect people. A parallel important difference between the two types of spaces is that nanotechnology cleanrooms typically have a slight over pressurization with respect to adjacent building spaces to prevent outside contamination from coming into the clean room while Biological Safety rooms are under pressurized to be sure that nothing gets out of the rooms. Design considerations for space organization, construction, ventilation requirements, space outfitting, furnishings, finishes, and fine details are also compared and contrasted. Finally, considerations for collocating cleanrooms and Biological Safety Laboratories are covered and the Birck Nanotechnology Center at Purdue University is used as an example of the trade-offs that must be made when collocating.

Introduction

This section compares the design of nanotechnology cleanrooms and bio-containment Biological Safety Level 3 and 4 (BSL-3 and BSL-4)[1] facilities. Can a bio-containment facility be successfully built in a nanotechnology cleanroom?[2] At first glance, the requirements seem much the same. Both types of spaces are highly controlled. Both types try to eliminate or limit contamination. Both spaces require users to wear protective gowns. But suiting up in a cleanroom is necessary to protect product, while suiting up in a bio-containment facility is necessary to protect people. A parallel important difference between the two types of spaces is

---

[1] http://en.wikipedia.org/wiki/Biosafety_level
[2] J.R. Weaver, "A Design for Combining Biological and Semiconductor Cleanrooms for Nanotechnology Research," JIEST 48(1), 75–82 (2005).

© Springer International Publishing Switzerland 2015
A. Soueid et al. (eds.), *Buildings for Advanced Technology*, Science Policy Reports,
DOI 10.1007/978-3-319-24892-9_10

that nanotechnology cleanrooms typically have a slight over pressurization with respect to adjacent building spaces to prevent outside contamination from coming into the clean room while Biological Safety rooms are under pressurized to be sure that nothing gets out of the rooms.

Design Challenges

Similar Space Organization

The two types of facilities have similar sounding organizational plans but they differ in function. Many cleanroom facilities have been built for various purposes (even catsup is bottled in cleanrooms). BSL-3 facilities are usually smaller, but just as plentiful. Facilities that qualify as a BSL-4 facility are few, and require additional levels of protection.

Both types of facilities—nanotechnology cleanrooms and BSL-4 laboratories—function best when constructed with an interstitial level above dedicated to mechanical and electrical operations. This interstitial space allows mechanics to move freely above the work area without intruding into the space, and without wearing protective clothing. Below the cleanroom or bio-containment facility, space is required to house specialty equipment and support functions such as waste handling.

A nanotechnology cleanroom facility usually has a perforated floor with return air plenums, specialty gas cylinders, vacuum pumps, waste handling equipment, and similar functions located below on the lower level. Cleanrooms with a width of up to 6 m can function and still have laminar air flow with a solid floor and low return air grilles on the perimeter walls. This is most effective when liquid waste control is not an issue.

Bio-containment facilities usually have a solid non-pervious floor with waste handling equipment and waste decontamination tanks on the lower level. BSL-3 facilities do not require the lower level since waste decontamination is not required.

Similar Construction

The two project types use similar construction for gown-up areas, but they serve different functions: nanotechnology cleanrooms require gowning for protection of the product, while bio-containment laboratories require gowning for protection of the operator.

Both facilities require filtered air. Cleanrooms require supply side filtration to increase the cleanliness and reduce particle size for the process being performed. The exhaust side requires filters or air burners to prevent particulates from escaping into the atmosphere. Containment facilities require filters on the supply side to prevent foreign particles from contaminating the air and exhaust side filters to prevent hazardous material and organisms from escaping.

Both building types require hard interior surfaces to allow for easy cleaning and removal of contaminants. They both require a sealed environment to prevent infiltration of unwanted material and outward flow of harmful contaminants. Cleanrooms are sealed from incoming contaminants; while containment facilities prevent contaminants from escaping.

Both require waste removal. In cleanrooms, filters and burn boxes are applied to exhaust air, and retention and dilution tanks are used for waste water. BSL-4 facilities require filters on exhaust systems and retention/kill tanks for liquid waste before it can be introduced into the local sewer systems. BSL-3 and BSL-4 biocontainment facilities require all contaminated and radioactive waste be autoclaved before disposal.

Ventilation Requirements

Nanotechnology Cleanroom

Supply air for a nanotechnology-cleanroom requires a highly filtered system with incoming filters, and high efficiency filters located to process the recirculated portion of supply air. Supply air has a high rate of recirculation through the high efficiency filters to provide the cleanliness specified in the Class rating: ISO Class 5/Federal Standard Class 100 (M3.5), ISO Class 4/Federal Standard Class 10 (M2.5), etc. The supply air is directed as vertical laminar flow, usually from ceiling to floor, through either the floor or wall grilles installed close to the floor to return air to the overhead recirculation filters. Fresh, filtered air is introduced to meet code occupancy requirements and to replace the operating air for exhausted devices, such as fume hoods, solvent hoods, vacuum pumps, etc. This usually comprises only a small amount of the total recirculated air in the occupied space. All of the equipment has ducted exhaust connections. The major portion of the ventilation air is recirculated. A ISO Class 5/Federal Standard Class 100 (M3.5) with a ±0.1 °C controlled space has an air change rate in the vicinity of 120 air changes per hour. The facility is designed for positive air pressure differential relative to the exterior air pressure (i.e., air leakage flows out from the facility).

Biocontainment Facility

Supply air requires a highly filtered system. The major difference between a nanotechnology cleanroom and a bio-containment lab is that in a containment lab all of the air is 100 % exhausted, none is recirculated. This is referred to as "once-through" air. The amount of air is determined by heat gain and ventilation requirements of the space and the cooling capacity of the supply air, coupled with the amount of air exhausted from hoods and other apparatuses. The amount of air is usually around 0.6–0.9 m^3/m^2 and about 6 air changes per hour. All containment labs have supply and exhaust duct connections and all hazardous equipment—like fume hoods—are ducted to the exterior. The facility is designed for negative pressure differential relative to the adjacent spaces within the building.

Space Outfitting

A nanotechnology-cleanroom is equipped with "tools" and hoods. The clean or loading portion of an apparatus is usually in the laminar flow of ultra-clean air. The exhaust and service end of this same apparatus is located in the dirtier return air plenum. This allows the filters and decontamination apparatus to be serviced from outside of the cleanroom, by people not required to be gowned. Biocontainment facilities also have "tools" and hoods but of a different type. Its tools consist of bench-top investigative equipment and floor-standing support devices like centrifuges, incubators, and ultra-cold freezers. All of the apparatuses are located inside the containment laboratory along with the investigators. Hazardous work is accomplished in Biological Safety Cabinets (BSC)[3] that are ducted when required. In BSL-4 facilities, the investigator is in a contained air suit, and the hazardous materials and equipment are located in ducted enclosures.

The primary barrier in bio-containment is the BSC, designed to the latest requirements of the NSF/ANSI Standard 49,[4] and upgraded every 5 years. Cabinet selection is based on research and production needs. They start at Type A (A1 and A2) for routine microbiological work. Type B (B1 and B2) for containment and direct removal of volatile toxic gasses and fumes used in conjunction with biological research. Used primarily in BSL-4 laboratories, glove box BSCs are totally enclosed and ducted.

Furnishings, Finishes, Fine Details

In both project types, the selection of laboratory furniture is important. It needs to be movable, easily cleaned, and adaptable. More laboratory furniture is required in the bio-containment facility because it requires laboratory-type workstations. There are a number of successful furniture solutions for BSL facilities. The cabinetwork should be easily maintained and durable, made either of steel or stainless steel. Laboratory furniture in bio facilities includes sit-down workstations and stand-up equipment benches. In most cases, the process "tools" in nanotechnology cleanrooms occupy most of the floor space with lab furniture used as support surfaces.

---

[3] Primary Containment for Biohazards: Selection, Installation and Use of Biological Safety Cabinets. 3rd Edition 2007 CDC.

[4] http://www.nsf.org/business/biosafety_cabinetry/index.asp; NSF International is a public health and safety company, not to be confused with the National Science Foundation.

Construction of either type of facility requires strict attention be paid to the fine details. A bio-containment facility construction drawing detailer can learn a lot from the cleanroom construction industry. This industry has solved many of the fine details, including installing and sealing light fixtures from the interstitial floor, sealing conduits, penetrations and doors, and avoiding construction cracks. Cleanroom designers also understand the science of pressure testing the environment, including the need for magnehelic gauges to indicate pressure differentials between spaces.

The need to properly filter exhaust air is very important in bio-containment facilities. Care must be taken to design the system to allow for redundant filter systems in order to keep the facility operational when the other is being serviced. Even vent pipes from the lab waste systems require a redundant filter system.

Colocated Cleanroom/Bio-containment

It is difficult to construct a bio-containment facility in the middle of a nanotechnology cleanroom. If for some reason it is required, it cannot be accomplished without a number of air locks, internal garment change areas and innovative mechanical designs to prevent contamination crossover to either environment. The facilities should be designed as separate entities, and connected only if necessary. Before attempting to design connected spaces, a material and people work flow diagram should be developed to organize all similar functions in one environment, and then move to the other environment for continuation of the investigative or production process. Products can be passed through air locks between different environments or packages sealed for transport to the other environment. Also, separate entrances to the bio-containment facility and nanotechnology cleanroom are highly recommended.

The Birck Nanotechnology Center at Purdue University provides an example of the trade-offs that must be made when collocating. It was decided in the planning phase that the biocleanroom would be a bacteria-free zone, confining active organism work to biosafety cabinets. It was also determined that BL-2 would be the highest pathogen level used in the facility. These concepts provided the basis for the biocleanroom design. Biocleanroom air handling was segregated from that of the nanofabrication cleanroom, i.e., there is no air interchange between the cleanrooms. Unlike the nanofabrication cleanroom, however, the biocleanroom utilizes a low sidewall air return rather than through-the-floor airflow. Solid floor tiles were used instead of perforated tiles, with welded-seam sheet-vinyl flooring placed over the tiles. This flooring was coved at the walls and joined to the wall system. The wall system was specially designed for pharmaceutical-grade cleanrooms and minimizes locations that allow bacteria growth. Terminal filters are integral to the ceiling system, which coves to the wall system.

Air-return chases were provided, but unlike the large chases used in the nanofabrication cleanroom, the chases are narrow and provide only sufficient room for airflow. The air returns were lined with a gently curved stainless steel liner to three feet above the wall opening, making the entire system cleanable with bactericidal agents. To minimize bio-entrapment areas, only wall penetrations were allowed for utility entry; no floor penetrations were allowed. Cores were intermingled with the chases to allow utilities to enter the biocleanroom from the sub-fab and under-floor area. These cores, although not part of the clean air path, bring utilities through the floor, allowing utilities to enter the biocleanroom through the chase-core wall. Utilities were run inside the air-return openings and connected directly to the equipment. The biocleanroom entry and exit facilities differ significantly from those of the fabrication cleanroom. A once-through disposable garment system is used for bacteria control. This system allows for a smaller gowning room with no need for staging facilities for previously worn garments. The less-stringent ISO Class 5 facility eliminates the need for pregowning and allows the use of a single air shower between the gowning room and biocleanroom. The exit path is distinct from the entry path and provides for garment disposal upon exiting the biocleanroom.

Because of the differences in gowning systems and protocols between the biocleanroom and the nanofabrication cleanroom, no direct personnel access is allowed between the two facilities. However, it is critical that devices can be moved between the two cleanrooms. For non-critical items, e.g., devices that have not been exposed to biological species, a conventional pass-through is provided. For safety reasons, an ultraviolet light is used in the pass-through to ensure that there are no viable species on the item. For critical items, a specially designed glovebox is used. This glovebox has glove ports in both the nanofabrication cleanroom and the biocleanroom and entry/exit ports in both cleanrooms. This design allows a researcher in one cleanroom to transfer an item into the other cleanroom, performing a decontamination or encapsulation process in the glovebox. This process provides versatility to researchers performing device development across both cleanroom facilities. Figure 10.1 provides a top view of this "double glovebox" arrangement.

Fig. 10.1 Schematic illustrating a glovebox arrangement that permits the transfer of samples into/out of a traditional cleanroom (from Weaver, see footnote 2)

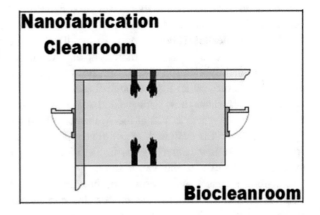

Acknowledgement Thanks to Mike Somin and John Weaver who synthesized material from several sources for the initial draft.

Bibliography

D.M. Carlberg, *Cleanroom Microbiology for the Non-microbiologist* (CRC, Boca Raton, FL, 2005). ISBN 0-8493-1996-X

D.S. Ensor, A.M. Dixon, *Environmental Monitoring for Cleanrooms and Controlled Environments*. Drugs and the Pharmaceutical Sciences (Informa, New York, 2006). ISBN 978-0-8247-2359-0

ISO 14698-1:2003, Cleanrooms and associated controlled environments—Biocontamination control—Part 1: General principles and methods by ISO/TC 209

B. Ljungqvist, B. Reinmuller, *Practical Safety Ventilation in Pharmaceutical and Biotech Cleanrooms* (PDA/DHI, Bethesda, MD/River Grove, IL, 2006)

Multiple publications are listed by The Pharmaceutical & Healthcare Sciences Society, http://www.phss.co.uk/?cart=yes&do=cat&cid=4

The Baker Company, Introduction to Biological Safety Cabinets (2010), http://www.bakerco.com/intro-to-biological-safety-cabinets.html

University of Pennsylvania, Clean Benches vs. Biosafety Cabinets (6 November 2009), http://www.ehrs.upenn.edu/programs/bio/biologicalsafetycabinets/cleanbenches.html

U.S. Centers for Disease Control and Prevention; U.S. National Institutes of Health (2000), http://www.cdc.gov/od/ohs/pdffiles/BSC-3.pdf

Chapter 11
Case Studies and Building Statistics

Abstract This chapter is divided into two parts. Part A gives detailed case studies of four completed facilities for research at the nanoscale; each is designed to technical criteria specific to the particular science program conducted at the facility. Part B provides short building statistics summaries of a number of completed facilities for nanotechnology R&D, including metrics and design criteria for the sake of comparison.

This chapter is divided into two parts:

- **Part A: Case Studies**
  This part consists of detailed case studies of four completed facilities for research at the nanoscale; each is designed to technical criteria specific to the particular science program conducted at the facility.
- **Part B: Building Statistics**
  This part consists of short summaries of a number of completed facilities for nanotechnology R&D, including metrics and design criteria for the sake of comparison.

Part A: Case Studies

Introduction

State-of-the-art laboratories built not too long ago are now maxed to the limit of their capability to provide a viable noise environment, making it extremely difficult to accommodate new research. Scientists are finding themselves spending time improving the physical environment, diverting valuable talent and time away from research. The most economical fix is to introduce self-contained mini labs to improve environmental characteristics around a specific experiment or piece of equipment. However, conflicting environmental criteria demand an increasingly complex infrastructure. Many institutions are finding that it is necessary to design and construct new facilities to meet the evolving environmental standards.

© Springer International Publishing Switzerland 2015
A. Soueid et al. (eds.), *Buildings for Advanced Technology*, Science Policy Reports,
DOI 10.1007/978-3-319-24892-9_11

The case studies presented here include differing building types and different approaches to meeting the need for precisely controlled research environments. The case studies include the following:

- *Advanced Measurement Laboratory*
  National Institute of Standards and Technology (NIST)
- *The National Institute for Nanotechnology*
  University of Alberta, Canada
- *Advanced Microscopy Laboratory*
  Oak Ridge National Laboratory
- Generic Analytical Instrumentation Facility
- Generic 300 mm Wafer Fabrication Facility

Case Study 1: NIST Advanced Measurement Laboratory

Introduction

For more than a century, the National Institute of Standards and Technology—an agency of the U.S. Department of Commerce—has developed the measurements and standards necessary for the United States to excel in technological innovation. In a global economy, completion of the Advanced Measurement Laboratory (AML) allows NIST to offer the U.S. science and industry the ability to accurately measure to new more precise, previously unimaginable standards.

Advances in science are now at subatomic and nanometer scales. Research at this level is highly affected by the surrounding environment, creating unprecedented need for environmental stability in the laboratory. Completed in 2004, the AML has few—if any—equals among the world's research facilities. It offers an unprecedented combination of features designed to virtually eliminate environmental interferences that undermine research at the cutting edge of measurement science and technology.

This \$235 million, 49,843-m^2 (536,500-$ft.^2$) facility features five separate wings with stringent environmental controls on particulate matter, temperature, humidity, vibration and electromagnetic interference allowing for key twenty-first century research. This new research and technology includes nanotechnology, semiconductors, biotechnology, advanced materials, quantum computing, and advanced manufacturing.

AML Design

Project Site

Built in the early 1960s, the NIST Gaithersburg campus buildings are somewhat diverse in age, function, and appearance. However, they are somewhat cohesive in the visual language of landscape, exterior building materials, and scale. In addition, physical links between facilities reinforce circulation patterns on campus. Part of the challenge was to design the AML to complement and enhance those aspects of the existing buildings which give character and operational coherency to the entire campus.

The AML site is located adjacent to the southern terminus of the main NIST circulation spine connecting a series of research facilities to the NIST Administration building on the north end of the campus. A two-level pedestrian link connects the AML to its closest neighbors (Buildings 220 and 221) where scientists work in similar fields of research as the AML occupants. The location of the AML creates an opportunity for collaboration and interaction among researchers, bringing similar groups and divisions together into one larger advanced technology complex.

In addition to its proximity to existing research facilities, one of the primary reasons for selecting the site for the AML was the ambient vibration levels at this particular location. The design team performed several 24-h site vibration measurements at different locations on the NIST campus. Because the ambient site vibration characteristics are critical to the function of the laboratories, this site was chosen over more prominent locations on campus.

Other factors were considered in selecting the site, including aesthetic concerns such as image, as well as practical concerns including land use and utility capacity. Underground utilities surround the site, allowing the AML to tap into existing campus distribution systems for water, sewer, steam, chilled water and power.

Building Layout

The AML is comprised of five distinct wings, each with a specific function. Two wings contain metrology laboratories, two contain instrument laboratories, and the fifth houses a nanotechnology fabrication facility. Wings are staggered along the circulation spine in a layout similar to other laboratory buildings on campus. A large light-filled atrium along the circulation spine provides organizational focus and physically links the five components of the program. While reinforcing the organizational arrangement of the AML wings, the atrium also serves as the southern terminus or destination of the campus circulation spine, and as a major entry portal into the complex.

The two-story atrium hub provides easy access to each of the wings as well as open views to the courtyard. This space is also designed to be a "celebration of science." Exhibit cases are integrated into the interior wall construction, promoting display of scientists' latest discoveries. The floors, ceilings, and walls are designed with patterns, colors and textures that subtly express scientific diagrams such as energy in wave and particle form, and the layering of materials in semiconductors.

Directly connected to the atrium are three of the AML wings, the two instrument laboratories and the nanofabrication facility. The instrument buildings are designed to respect the scale of the existing campus buildings, while the nanofabrication facility punctuates the state-of-the-art research that takes place inside. The latter is a highly specialized cleanroom facility supporting research conducted throughout the campus.

The site slopes gently from the northeast to the southwest at approximately a 6 % grade. The multi-level nanofabrication facility is located at the site's lowest point to allow its sub-fab level easier access to the loading dock and utility tunnel. A naturally screened equipment yard houses the emergency generator and gas tanks that support cleanroom operations.

The metrology laboratories, which require extreme control of temperature and vibration, are located below grade just to the north of the instrument laboratories and are not visible from ground level. While hidden from view, landscape treatments and stair connections above grade offer visual cues to the presence of the metrology facilities.

Organization and Design

The instrument laboratory wings express aspects of their operation in the architectural language of the structure. Two distinct major functional elements, laboratories and offices, are contained in the wings, with the laboratories located in the center. Mechanical ventilation devices articulate the large three-level laboratory blocks, clearly referencing their technological function. Offices are housed in attached two-level structures flanking the north and south ends of each laboratory block. Smaller in scale, the office facades are punctuated by window openings allowing natural light into the interior. The perimeter elevation is pierced by shared break spaces sheathed in glass projecting from the building. These spaces act as nodes of activity within the circulation system of the building. The architecture of the instrument wings expresses its mission, both for the people and the technologies that it supports.

The wing housing the nanofabrication facility and attached atrium employs machine-like architectural articulation in alignment with its highly specialized technological function. Linear visitors' galleries are located behind inclined windows with suspended metal canopies above, running the length of the building. At the rear, large service galleys with dramatically inclined surfaces are cloaked in metal louvers.

Image

The AML incorporates design elements from the existing campus. Incorporating the predominant material of the existing surrounding buildings—a buff-colored brick—the design of the AML provides visual continuity and enhances its integration into the campus. The brick is used to enclose the smaller-scale office components in the instrument laboratory buildings, providing a traditional character and human-scale to the "people" side of the facility. Laboratory components, including the nanotechnology fabrication facility, employ manufactured materials such as formed metal panels, to express the powerful technologies within. A departure from the contextual language of the campus, the facades express the leading-edge science capabilities of these new facilities.

Exterior glazing is used to highlight portals of entry, important circulation intersections, and critical nodes of interaction. In all, the palette of materials expresses both continuity and connection to the historic development of the NIST campus and the visionary purpose and technological advancement embodied in the AML.

Critical Design Criteria

As nanotechnology research compels the scientific world to explore new uncharted territories, scientists are increasingly demanding more stable research environments. NIST scientists working in the AML are manipulating matter at the atomic and molecular scales in order to obtain materials and systems with significantly improved properties. Working to push beyond the limits of today's advanced technologies, NIST AML researchers crave environmental stability in their laboratories. Even tiny variations in environmental conditions—a hundredth of a degree change in temperature, vibrations from local traffic, a flutter in electrical current—can plunge the results of the most carefully designed experiment into ambiguity.

The AML includes sophisticated spaces specifically designed for making extremely small measurements needed for the design of more durable nanoscale-sized gears and other devices—nanomachinery of the future. The physical environments needed to conduct such research increasingly impose strenuous demands on facilities. These demands include high levels of control of environmental criteria such as[1]:

- Temperature control
- Humidity control
- Vibration isolation
- Acoustic isolation

---

[1] A Soueid, H Amick, T Zsirai, "Environmental Challenges of the NIST Advanced Measurement Laboratory," Proceedings of SPIE Conference 5933: Buildings for Nanoscale Research and Beyond (Aug 2005).

- Air cleanliness from particulate matter
- Control of biological contaminants
- Electromagnetic interference (EMI)
- Radio frequency interference (RFI)
- Good quality electrical power

Many of these criteria have been individually achieved at similar facilities around the world. However, the greatest design challenge at the AML was to achieve combinations of various criteria that often conflict with one another. In attempting to control air cleanliness, greater volumes of air are required. The movement of air, along with the larger mechanical equipment needed, results in a greater potential for acoustical noise and mechanical vibration interferences. Likewise, in attempting to control slab temperature by way of liquid systems embedded in the slab, the movement of fluids induces vibration. Alternately, the use of electric heating elements in the slabs creates electromagnetic fields.

The design process included several research projects to determine the criteria and develop the right design solutions for the AML. In order to incorporate lessons learned from other similar projects, the design team and NIST representatives toured similar facilities around the U.S. and Europe. Furthermore, architectural and engineering solutions developed for the AML were tested early in the design process in fully functional mockups. In addition to mockups built for a typical laboratory and office, a Vibration Isolation Research Project (VIRP) was built to test the design concept for vibration isolation. A Temperature Control Research Project (TCRP) was also built to test the design of the HVAC system for the 48 high-accuracy control laboratories. These mockups proved beneficial for NIST researchers, the architecture and engineering team, as well as potential manufacturers of equipment to be installed at the AML. Lessons learned from these mockups were incorporated into the design of the building long before construction started. During construction, additional mockups were built to assist in coordinating among construction trades.

Vibration Isolation

Vibration levels can disturb the results of sensitive measurements. The AML was designed to achieve the "NIST-A" criteria in most laboratories (0.025 μm displacement for $1 < f < 20$ Hz; 3 μm/s velocity for $20 < f < 100$ Hz). Research laboratories are located on or below grade, the best possible location with respect to vibration. The below-grade placement of the metrology labs enables a greater degree of vibration control by avoiding vibration waveforms that travel at and near the ground surface, as well as avoiding vibration induced by wind on the structure of the buildings.

Where more stringent vibration controls are necessary, special isolation slabs were constructed and supported on air springs. The AML incorporates air springs technology with special active and passive controls that provide proper isolation to meet the most stringent vibration criteria "NIST A-1" (velocity of 6 μm/s for $f < 5$ Hz and 0.75 μm/s for $5 < f < 100$ Hz). The isolation slab design is flexible with replaceable air springs, allowing scientists to customize laboratories as needs or technologies change. The most sensitive areas in the metrology wings are located about 40 ft. below ground level.

Temperature and Humidity Control

Temperature fluctuation within a room can compromise the accuracy of very sensitive measurements. Eliminating temperature fluctuations has long been an elusive goal for metrology laboratories. The baseline temperature of the majority of the AML laboratories is controlled to within ± 0.25 °C. In the 48 high-accuracy control laboratories in the metrology wings, the temperature is controlled to within ± 0.1 or ± 0.01 °C. Humidity control provides variations of no more than 1 % relative humidity in specialized areas and 5 % relative humidity throughout the rest of the facility.

Providing temperature control to 1/100th of a degree Celsius required careful design considerations as the volume of air needed and the various stages of treatment mandated a large volume of space within the building. Control systems for this level of performance did not exist during the early design process. The final concept was specially designed and manufactured from existing components through a collaborative effort among NIST scientists, HDR architects and engineers, and four different HVAC controls manufacturers.

Clean Electrical Power

In addition to precise control of temperature, humidity, and vibration, the labs are serviced by clean electrical power and varying degrees of EMI shielding. Clean power is important for the accuracy of the instrumentation utilized in the AML. Providing high-quality power involves a difficult compromise. The quality of power improves as the distance between the conditioned source and lab instrument is minimized. However, power conditioning equipment creates electromagnetic interference, heat, noise, and dust, all major problems in these laboratories. Five different approaches were considered to determine the optimal balance of these competing criteria. Conditioned power to the labs is provided by on-line static uninterruptible power supply (UPS) units. UPS provides outage and transient protection as well as protection from noise, voltage sags and swells, and over and under voltage conditions.

Particulate Control

Because dust or other stray particles can foul measurements on atomic-scale devices, the air inside much of the AML building is filtered through HEPA (High-Efficiency Particulate Air) technology. The nanofabrication facility is designed to have no more than 3.5 particles (at 0.5 μm) per liter of air (ISO Class 5/FED STD 209E Class 100/M 3.5) where air flows downward through the clean space at an average velocity between 0.2 and 0.5 m/s. The facility is designed to be upgradeable to ISO Class 4/Class 10/M 2.5. Some spaces within the instrument and metrology blocks also require cleanroom conditions. These rooms have ceiling-mounted HEPA filters with return air through low wall registers or a raised floor.

State-of-the-Art Laboratories

Over 100 different types of laboratories were identified during the design phase of the AML. Classifying the spaces into a few general categories helped to establish modules. This modular approach to the organization of lab space and distribution of utilities can be readily modified in anticipation of changing research efforts in the future.

The standard lab module is 6.9 m long by 3.6 m wide. The primary lab blocks are arranged linearly with a service corridor flanking one side and a circulation corridor flanking the other. The service galleys contain floor areas designated to house scientific support equipment. The floor of the service galley is isolated from the floor of the laboratories to make the segregation of rotating equipment more efficient. The upper level of the services galleys contains the distribution of lab utilities. An accessible ceiling space above the labs provides access to the air supply and return ducts, terminal boxes and coils. The ceiling system can support walking loads and the removable panels facilitate changes in air supply, exhaust, or lighting in the labs.

Specialty areas within the AML include 48 precision temperature control laboratories and 18 extremely low-vibration laboratories. The AML also includes provisions for future retrofits of specialty areas.

In addition to providing state-of-the-art, flexible research environments, the design incorporates laboratory planning concepts such as modular lab dimensioning; hierarchical zoning; separation of service galleys from staff/public circulation corridors; interstitial service zones; dedicated zones for services/utilities and modular provisions for future service and space upgrades. Natural day lighting, energy conservation and recycling are also incorporated into the building design and operation of the AML.

The Nanofabrication Facility

The nanofabrication wing is designed as a multi-level cleanroom with a full sub-fab below the clean research floor. An interstitial space dedicated to mechanical distribution separates the clean floor from the mechanical penthouse. The waffle slab and its supporting structure below the cleanroom floor were specially designed to provide the vibration absorption characteristics of a slab-on-grade condition.

A visitor gallery on the clean floor level surrounds the wing, permitting views into this restricted environment. This allows for visitors to view operations inside the nanofabrication facility without having to go through a gowning and de-gowning procedure and without disrupting ongoing research. The bay and chase cleanroom concept is designed to be ISO Class 5 upgradeable to ISO Class 4. The cleanroom wing consists of a series of bays and chases designed in 8-m-long lab modules and arranged along either side of a central clean aisle. Access to this corridor is through a gowning area and an airlock vestibule. Service corridors flank the modules for materials handling and utilities support services.

Instrument Labs East and West

The instrument lab blocks house 187 lab modules. The typical temperature criterion for most labs in these sections is temperature $= 20$ °C controlled to ± 0.25°C. However, several labs are designed to reach a temperature control of ± 0.1 °C. Similarly, particulate cleanliness for most labs is ISO Class 7 with several labs modified to reach ISO Class 6 and 5.

Metrology Labs East and West

The two metrology lab blocks house 151 lab modules. Both of these lab blocks are located approximately 12 m (40 ft.) below grade, providing additional shielding from vibration sources and temperature fluctuations.

The metrology blocks house two types of labs: "quiet" metrology labs which primarily consist of measurement labs and "rotating" or "dynamic" metrology labs where moving equipment is used. The design separates the two functional types of space into separate building blocks. Rotating metrology labs are located on the west side of the site and quiet metrology labs are in the optimal location for vibration isolation on the east side of the site.

Most of the high-accuracy laboratories requiring ± 0.10 or 0.01 °C temperature control are located in the metrology wings of the AML. These labs are designed as a room-within-a-room. Double-insulated wall construction provides an annular space which acts as a buffer and allows for the circulation or return air around the critical portion of the laboratory. The floor of the space is raised, providing a 1-m space below for air flow. The room walls consist of 100-mm-thick insulated metal panels forming the inner and outer walls and ceiling.

NIST: A Vibration Criteria

The NIST-A criterion was developed in the early 1990s for the AML at NIST. There are others, such as the VC criteria developed in the early 1980s[2] and discussed at length in IEST-RP-CC012. The NIST-A criterion was developed for metrology, but has gained popularity within the nanotechnology community.[3]

The NIST-A criterion takes the form of a one-third octave band velocity spectrum. It refers to vibration as measured in the vertical and two orthogonal horizontal directions, and is applied to each direction separately.

For environments that are continuous and steady-state in time, the criterion applies to the "linear average" of data samples acquired over an adequate time period. In instances where the environment is impacted by occasional disturbances such as vehicular movements, "stage" movements (in tools), passing trains, etc., these may be evaluated in the "peak hold" or "maximum RMS" mode of the measuring system. If the disturbing event is long enough (i.e., "quasi-static", or steady-state during the averaging time) the linear average mode should be used. The importance attributed to these occasional events will depend on how frequently they occur and other parameters relating to the vibration-sensitive process.

The main elements of the criterion are:

1. Vibration is expressed in terms of its root-mean-square (RMS) velocity (as opposed to displacement or acceleration)
2. The use of a proportional bandwidth (the bandwidth of the one-third octave is 23 % of the band center frequency) as opposed to a fixed bandwidth is justified on the basis of a conservative view of the internal damping of typical equipment components. Experience shows that in most environments where adequate layout and isolation of electrical and mechanical equipment has been provided, the vibration is dominated by broadband (random) energy rather than tonal (periodic) energy.

For a floor or site to comply with a particular equipment category, the measured one-third octave band velocity spectrum must lie below the appropriate criterion curve of Fig. 11.1. It is generally accepted that vibration measurements are accurate and repeatable only within about 1 or 2 dB (12 or 26 %), so an overly strict

---

[2] Vibration Criterion (VC) http://www.newport.com/Environmental-Vibration-Criteria/168090/1033/content.aspx; E. E. Ungar and C.G. Gordon, "Vibration Challenges in Microelectronics Manufacturing," *Shock and Vibration Bulletin*, 53(I):51–58 (May 1983); and C. G. Gordon and E. E. Ungar, "Vibration Criteria for Microelectronics Manufacturing Equipment," *Proceedings of Inter-Noise 83*, pp. 487–490 (July 1983).

[3] H. Amick, M. Gendreau, and C.G. Gordon, "Facility Vibration Issues for Nanotechnology Research," *Proc Symp. on Nano Device Technology* (May 2002).

Fig. 11.1 Generic
Vibration Criterion (NIST-
A) Curve for critical areas
in nanotechnology
facilities—showing also
several of the VC criteria
for reference

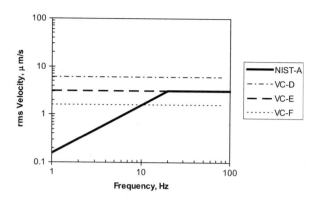

interpretation of a comparison to the criteria is not encouraged. For instance, a
measured value of 51 μm/s versus one of 49 μm/s, when compared to a criterion of
50 μm/s, lies within the range of inaccuracy—less than 1 dB—with respect to the
criterion. It may be argued that both of them meet the criterion for measurement
accuracy.

The NIST-A criterion is identical to VC-E at frequencies above 20 Hz, but
maintains a constant rms displacement amplitude at lesser frequencies to accommodate some of the ultra-high-precision metrology, probe and lithography equipment used in nanotechnology. This is a difficult criterion to meet at sites with
significant low-frequency vibration content.

Facility vibrations do not necessarily remain constant over extended periods of
time. Vibrations measured during construction may not reflect the contribution of
the mechanical systems in their operational state at building completion. Likewise,
vibrations at a few months beyond completion may include contributions from
user-installed equipment, and this contribution could change over time as equipment layouts change. (This variation has been called "maturation," and must be
considered a normal part of the aging process.[4]) It is important that a facility survey
be carried out at a time appropriate for characterization of the operational state of
interest. For example, one would not want to characterize the "as built" state using
measurements made either during construction or at 1 year after startup.[5]

In most instances it is recommended that the advice of a vibration consultant be
sought in selecting a design standard (Table 11.1).

---

[4] M. Gendreau and H. Amick, "Maturation of the Vibration and Noise Environments in Semiconductor Production Facilities," *Proc. ESTECH 2004*, 50th Annual Technical Meeting, Institute of
Environmental Sciences and Technology (IEST), Las Vegas, Nevada, April 28, 2004.

[5] H. Amick, M. Gendreau, and T. Xu, "On the Appropriate Timing for Facility Vibration Surveys,"
Semiconductor Fabtech, No. 25, March 2005, Cleanroom Section.

Table 11.1 Numerical definition of criterion curves shown in Fig. 11.1

Criterion	Definition
NIST-A	0.025 μm or 25 nm (1 μin.) between 1 and 20 Hz; 3.1 μm/s (125 μin./s) between 20 and 100 Hz
VC-D	6.25 μm/s (250 μin./s) between 1 and 80 Hz
VC-E	3.1 μm/s (125 μin./s) between 1 and 80 Hz
VC-F	1.6 μm/s (62.5 μin./s) between 1 and 80 Hz

Case Study 2: The National Institute for Nanotechnology, The University of Alberta

Introduction

Created as a partnership between the National Research Council, the University of Alberta and the Governments of Canada and Alberta, Canada's National Institute for Nanotechnology (NINT) combines the research and development strengths of the National Research Council of Canada (NRC) and the University of Alberta. This hybrid organization brings together scientists, engineers, entrepreneurs and graduate students from private and public sector organizations. Its current mission is to "to transform nanoscience ideas into novel, sustainable nanotechnology solutions with socioeconomic benefits for Canada and Alberta."

Currently located in a purpose-built facility on the University of Alberta campus, NINT houses numerous wet and dry labs, specialized electron microscopy and cleanroom spaces, and a research transfer facility that is home to more than a dozen spin-off and early market nanotechnology-enabled companies. In 2011 a million-Canadian-dollar expansion and renovation to its characterization suite was completed to house the Hitachi Electron Microscopy Products Centre and its three new electron microscopes.

NINT Project Summary

Personnel and Facilities

- 100 institute staff
- 100 or more collaborating researchers
- 150–200 graduate students
- 23,588 m^2 (255,000 ft.^{2}) gross. building (CAD$51.5 million)
- Approx. CAD$65 million of research equipment
- Approx. CAD$15 million for annual operations
- Approx. CAD$5 million annually in industrial research collaborations

Program Summary (Table 11.2)

Table 11.2 NINT Building Areas

Component	Area (m^2)
Technical/research laboratory	
High-end characterization (STEM, AFM)	574
Cleanroom/nanofabrication	1777
Interdisciplinary generic labs (NRC)	3150
Interdisciplinary generic labs (engineering)	3150
Research staff facilities	
Research offices and workstations	1772
Seminar/conference facilities	687
Administrative/public/building support	
Management/administrative & support	273
Public facilities	191
General/common building support	397
Industry partners/incubator facilities	1904
Total program net area	13,875
Total gross building area (gross-up factor 1.70)	23,588
	(approx. 255,000 ft.^{2})

Research Program

1. Metabolomic Sensors
2. Hybrid Nanoscale Electronics
3. Energy Generation and Storage
4. Nano-Enabled Bio-materials

To accommodate the requirements of a nascent research program, challenges in the initial design and construction phase included site selection, and the design of cleanrooms, characterization suites and other specialized laboratories for a yet to be finalized tool set. In particular, process and characterization tools required stringently controlled environments. The NINT design team identified nine key facility design issues:

- Electromagnetic Interference (EMI)
- Vibration
- Temperature
- Air Flow
- Air Pressure
- Humidity

- Cleanliness
- Power Quality
- Ground

Overlaying these facility design issues was a modest project budget, constantly evolving research tools, research programs that will most likely change, and a need to have the new building constructed and operational. Flexibility, expediency, and fiscal prudence were seen early on in the design as prime drivers of the project.

A decision was made early in the project to assemble a design team comprised of a prime consultant who was knowledgeable, enthusiastic, innovative and accessible as well as other specialty consultants. Integral to the design team was a small group of "hands-on" personnel from the Faculty of Engineering at the University—the owner of the building—and the NRC, the tenant. A project manager and a construction manager, especially important for constructability issues, were also brought on board early. The project prime drivers and the nine key facility design issues were discussed in detail at the beginning of the project with all members of the design team in a series of workshops and meetings, and also on site visits to facilities in the U.S. and Europe.

NINT Phase 1 Project Summary

Canada's National Institute for Nanotechnology has now occupied its new building for many years. Changes and additions to the original plan include the completion of a research transfer facility on the fourth floor. The NINT Innovation Centre is home to between 12 and 15 spin-off or start-up companies and industrial research collaborations; each housed in a rental unit of combined office and lab space. The characterization suite has been expanded and modified to house three new electron microscopes in the Hitachi Electron Microscope Productions Centre (HEMiC). Most recently, a High Performance Computing facility came on-line. Today, NINT has more than $65 million in electron microscopy and instrumentation.

Case Study 3: Oak Ridge National Laboratory Advanced Microscopy Laboratory

The introduction of electron microscopes with ultra-high spatial and energy resolution has resulted in the necessity to construct special laboratories that provide the quiet environment needed for these sensitive instruments to achieve their design resolution performance. The Advanced Microscopy Laboratory (AML) at Oak Ridge National Laboratory is such a facility. It is a new 6900 ft.^{2} world-class microscopy laboratory built to house the latest generation electron microscopes that can resolve, for example, single isolated atoms, or columns of atoms in a crystal separated (in projection) by less than the diameter of a single atom.

These next-generation microscopes are perhaps the best detectors in the world for environmental disturbances such as floor vibrations, alternating magnetic fields, acoustic noise (e.g., microphonics), air flow, and changes in temperature and air pressure. Each of these disturbances can be minimized by using appropriate methods of construction of the laboratory. For example, heavy isolated slabs are generally used to attenuate vibrations from external as well as local sources. Magnetic fields are minimized by paying careful attention to building power wiring design, the location of electrical transformers and distribution panels, and the location of external field sources such as power lines or equipment in adjacent buildings. Microphonics are controlled by appropriate soundproofing design. The design and construction features of the AML that are important in ameliorating these environmental disturbances are described in greater detail in the following paragraphs.

The AML building is on a site adjacent to the High Temperature Materials Laboratory (HTML) at ORNL, and takes advantage of chilled water, compressed air, and clean power supplied from the HTML. The site and building have been designed to allow future expansion to twice its size to accommodate four additional instruments. The AML actually provides a pair of rooms for each microscope—an instrument room and an associated "control room" which contains the computers used to operate the microscope, and additional facilities for data analysis. This design effectively isolates an instrument from even the disturbing influence of human operators while critical imaging is being done. The laboratory also comprises a small sample preparation room, a lobby for visual displays, and a building office. All ancillary equipment (with the exception of the water chillers) is housed in a common service chase that has a foundation isolated from the instrument rooms. A structurally isolated mechanical building houses the HVAC equipment for the laboratory.

The foundation slabs and wall footings of the AML building were placed on 2.4 m (8 ft.) engineered fill separated into layers by geotextile fabric. Geotextile placement and fill placement and density were closely monitored. The building exterior is EIFS (Exterior Insulation and Finish Systems) construction.

Floor slabs in the instrument rooms are 0.3 m (1 ft.) thick reinforced concrete, and cover the full floor area of the room. Epoxy coated reinforcement bars (re-bars) were used to provide insulation and to minimize the possibility of magnetic fields caused by currents in the foundation. Re-bars were connected with PVC coated tie wires. Any nicks or scrapes in the reinforcement were touched up with epoxy coating prior to installation. This slab-on-grade construction was deemed suitable because the site is inherently very quiet and provides exceptionally low vibration levels in the instrument rooms. Also, this foundation design for instrument rooms in the adjacent HTML has proven to be effective for controlling vibrations in that building to the level of 1 μm/s, as required for the AML instruments.

The instrument rooms section was constructed using a "house-in-house" design philosophy, with a shell built from 0.3 m (12 in.) reinforced concrete masonry blocks. The roof height is 7.9 m (26 ft.) in the instrument bay section. The instrument and control rooms were built using 0.23 m (8 in.) reinforced concrete blocks. This approach reduces the pressure differential, acoustic vibrations, and low-frequency vibrations produced by the wind moving over the building roof. Masonry block voids were grout filled.

Acoustic noise in the instrument rooms has been minimized by the application of sound absorbing material on all four walls and the ceiling. Absorber/barrier blankets have been installed covering all wall and door surfaces within an instrument room. The primary ceiling height in the instrument rooms is 5.5 m (18 ft.). A dropped ceiling at the 4.0 m (13 ft.) level, consisting of egg-crate grid and two layers of porous duct liner material, has been installed as an added noise reduction feature. The dropped ceiling forms an air mixing plenum for the supply air, which enters through a pair of 0.3 m (12 in.) diameter porous ducts that extend the length of the room. Return air plenums were formed by sheetrock walls built out 0.25 m (10 in.) from two opposing concrete block walls, with the return intake at floor level along the full length of each wall.

A single 1.2 m × 1.2 m (4 ft. × 4 ft.) window comprising a pair of double-paned windows is provided in the wall between the instrument and control room. The control rooms have cloth-covered acoustic absorber panels on each wall to absorb noise from conversation and computer fans. With most of the heat-producing equipment located in the mechanical section, air supply to the instrument rooms can be minimal and is much more easily controlled to avoid sudden large fluctuations. The lower air flow also provides the advantage of much less chance of low-frequency (sub-sonic) "booming" in the supply ducts. Sudden internal building pressure changes are avoided by the use of airlocks. Typical temperature fluctuations are kept to ±0.1 °C per hour.

The separate, isolated mechanical section houses the AML utility equipment and electrical load center. The equipment for the HVAC system for the instrument rooms and the control rooms is in the mechanical section. Each of three air handling units were installed on their own isolated slabs. Also, a separate isolated slab was provided for the water chillers that supply cooling water to the four instrument rooms.

Variable frequency drives were used to provide tight temperature and airflow control. To avoid the temperature variations caused by cycling air conditioning, the instrument rooms are relatively large (5.2 m × 5.2 m × 5.5 m) (~17 ft. × 17 ft. × 18 ft.) and have only the minimum number of heat-producing sources. Ductwork, piping, and hangers were installed using electrically isolated joints. The ductwork includes sound attenuators and internal liners. To minimize any possibility for generating magnetic fields, dielectric unions were used for the water lines, compressed air lines, and sprinkler piping installations. These unions were installed every 3.0–4.6 m.

Several measures were used to reduce or eliminate the effect of the electrical services on the microscopes. The electrical load center was located as far as practical from the instruments. A single point grounding system was used. Each instrument has been provided with its own ground system. PVC conduit was used and was installed below the floor slab. Twisted-pair wire has been used throughout the facility rather than straight runs of wire. The possibility of magnetic fields caused by currents in the foundation has been minimized by the use of epoxy-coated rebar, tied together with plastic-coated wire.

Clean power is supplied to the AML from a 75 kVA motor/generator (MG) set that is located in the HTML mechanical building. The MG set has an uninterruptible power supply (UPS) capability to provide power in the event of a power failure. This gives the advantage that the noisy pumps, compressors, and the MG/UPS set are totally isolated from the AML.

Case Study 4: Generic Analytical Instrumentation Facility

Introduction

Materials for nanotechnology are imaged and chemically analyzed at the atomic level. Performing these experiments requires highly specialized analytical instrumentation. Simply purchasing these tools does not ensure that they will achieve their required capabilities. Great care must go into the design of a facility intended for these instruments in order to eliminate any environmental effects upon their performance.

In designing a facility for analytical instrumentation, there are a number of requirements that must be met. Some of these requirements are general, such as lighting, heating, ventilation, privacy and safety. Due to the sensitive nature of the instruments needed for characterization in nanotechnology, a number of special requirements also exist. These instruments are sensitive detectors of mechanical vibration, magnetic fields and electrical, thermal and pressure disturbances. The performance of an instrument is seriously degraded if the various ambient disturbances are not reduced to an adequately low level. It is easier to initially design a facility to eliminate these factors than to correct for them after installation of the instrument.

This case study will discuss the deleterious effects of ambient disturbances on the performance of high-end analytical instrumentation. Specific requirements for the installation of atomic resolution transmission electron microscopes (TEM) will be given, as these instruments are often some of the most sensitive installed in these facilities.

General

When designing a facility for general instrumentation, basic requirements for the room that will contain the instrument need to be met. The room must be of adequate size to contain the instrument, and access into the room must be such that the instrument can be fit through the door prior to installation. Lighting must provide adequate illumination, and heating and ventilation must create a comfortable work environment. The room that contains the instrument should be as isolated as possible within the building to facilitate privacy and minimize disturbances during delicate experiments. Finally, the facility must be designed to meet all safety codes.

Special

For highly sensitive analytical instruments, a number of special requirements must be met for the instrument to achieve its utmost potential. Some of these specifications include: acoustic noise, mechanical vibration, electromagnetic interference, electrical disturbances, and environmental changes. Any of these factors is enough to hamper the performance of an instrument.

Acoustic Noise

Vibrations due to noise can occur from different sources: those external to the facility and those internal. Proximity to roads or highways, airports and railway systems will lead to acoustic vibrations. Heavy machinery tends to be loud and creates noise. Neighboring laboratories may include noisy equipment such as vacuum pumps, and any facility will contain a number of people—prone to talking and transmitting their own acoustic signature. To minimize these external sources of acoustic vibrations, a number of steps can be taken. The room containing the instrument should have thick, heavily insulated walls and doors. Better yet, the most effective manner in which to dampen external noise is to design the instrument room as a room within a room. Having two sets of walls to the room, separated by a cushion of air, will eliminate most acoustic vibrations.

Eliminating external sources of acoustic noise is not a complete solution. There are a number of sources of noise located in the room with the instrument that must be dealt with. The air conditioning and ventilation system in the room can generate acoustic vibrations as can other cooling fans, electronic noise from power supplies, and human noise. The most obvious solution is to remove as many of these sources from within the room as possible. Power supplies and vacuum pumps can often be moved to an adjacent room or chase. For other sources of internal noise, an in-room dampening system, such as foam or baffles on the walls and ceilings will help.

Mechanical Vibrations

For TEMs, there are a large number of resonant frequencies of the instrument structure and components that can affect the overall performance of the instrument. Specimen holders that place a sample into the electron beam tend to resonate at higher frequencies, while the heavier lenses of the microscope itself tend to resonate at lower (<30 Hz) levels. Sources of mechanical vibrations can include proximity to roads, airports and railway systems, heavy machinery, elevators, ventilation and air conditioning systems, other laboratory equipment, and human-induced vibrations. Equipment associated with the instrument itself, such as the vacuum system, power supply system and cooling system can also act as sources for vibration.

The primary solution to eliminating mechanical vibrations is to find an adequate location for the instrument. If at all possible, the facility should be located away from highways, train stations or airports, and places where heavy machinery is constantly moving about. In addition to a good location for the facility, the location of the instrument within the facility is also important. Locate the instrument as far away as possible from loading docks, storerooms and stockyards, elevators and from rooms containing the equipment that drives them.

If possible, it is best to decouple the room containing the instrument from the rest of the building. This can be accomplished during initial design phases by having an isolated support pad for the instrument room. This requires that the instrument be located on the ground level of the building.

If it is not possible to remove these sources of mechanical vibrations, the final solution is to use a vibration dampening system. This can be done in two steps. If mechanical vibrations are being caused by other equipment in the facility, it may be possible to place that equipment on vibration dampening pads. The instrument itself can also be stabilized using a vibration dampening or isolation system.

Electromagnetic Interference

Lens systems and beam deflectors in electron microscopes are extremely sensitive to perturbations induced by external magnetic fields. Spectrometers used in conjunction with the microscope for acquiring chemical information about the specimen are similarly affected by electromagnetic interference. Sources of electromagnetic interference can include: electricity supplies (such as high voltage transmission lines and local substations); subways or other electrical trains; three-phase electrical supply within the building (if unbalanced, this causes a current within the neutral conductor); electric motors from other equipment (such as elevators, hoisting cranes or the building AC plant); other laboratory equipment (arc-welders and other arc-discharge equipment like coaters); lighting; and stray fields associated with computer monitors in the room.

For a straight conductor, the magnetic field generated is determined by Ampere's law:

$$B = 2I/d$$

B is the magnetic field in mG; I the current in mA; and d the distance from the conductor in meters. The direction of the generated magnetic field follows the right-hand rule. For TEMs, a 0.3 mG rms magnetic field can be visibly detected in a scanning transmission electron microscope (STEM) image. At 1 m away from a wire carrying 0.5 mA of current, a 1 mG field will be produced—enough to degrade the 0.3 nm imaging performance of a STEM.

As with acoustic noise and mechanical vibrations, the best solution to electromagnetic interference is to design the facility to avoid magnetic fields. Once a facility is complete, it is much more difficult (and expensive) to try and reroute wiring to eliminate fields. Keep the facility away from high-tension wires, power stations, and subway lines. Within the building, it is best to keep the instrument away from internal mains lines, elevators, AC generation equipment, and other electrical laboratory equipment.

Electromagnetic field cancellation systems can only correct for one position on the microscope. This means that the operator may have to choose between correcting fields for the illumination system of the microscope, the imaging system of the microscope, or a spectrometer. These cancellation systems thus work best in large rooms, where the correcting field is most uniform and the external field is uniform across the entire TEM column.

Environmental Changes

Ambient environmental conditions in the facility housing an instrument can have as detrimental an effect on performance as any of the previously mentioned causes. Important factors can include the airflow within the instrument room, temperature control and pressure fluctuations.

Airflow within a room can be a source of both mechanical vibrations and thermal fluctuations. For 0.2 nm resolution electron microscopy, airflow across the microscope column must be less than 4.5 m/min. Common sources of air movement in a microscope room may include: the room air conditioning system, an HVAC vent, a person walking, a door opening, or a diffuser vent. Directing airflow vents away from the column and to the outer walls of the room will help eliminate air flow across a column.

For a typical forced air temperature control system, the heat load in the equipment room determines the air flow rate:

$$q = mC_p\Delta T$$

In this equation, q is the heat load, m the mass of air moved, C_p the specific heat of air and ΔT the temperature gradient between supply and return. In a $3 \times 4 \times 3$ m^3 room with a heat load of 4 kW and 100 air changes/hour, the temperature gradient would be on the order of 4 °C. This level of thermal instability leads to thermal drift of the specimen holder in the microscope, as well as thermal drift of the lens systems themselves. Furthermore, this is a lot of airflow across the column.

Systems other than forced air temperature control, such as radiant heat can greatly improve thermal stability to levels below 0.1 °C. Other solutions include construction of a "gazebo" around the microscope to prevent airflow and wrapping the microscope column with neoprene to eliminate thermal fluctuations associated with airflow.

Remote Instrument Operation

In each case outlined above, one source of instrument degradation is related to the instrument operator. An operator at the microscope creates air flow walking around the room, induces acoustic vibrations by talking and moving about, and creates local thermal fluctuations due to body temperature. Furthermore, thermal fluctuations are generated by the need to keep the work environment comfortable for a user. Microscope lens systems are cooled to around 15 °C. Typical room temperature is 23 °C. This temperature difference leads to both the cooling system of the microscope and the temperature control system of the room needing to compensate.

One solution is emerging to eliminate all of these issues—remote operation. Most electron microscopes are now controlled by PCs, and some manufacturers have the ability to operate these control PCs, and thus the instrument, remotely. Thus, the operator can be removed from the local instrument environment, improving overall stability and performance.

Case Study 5: Generic 300 mm Wafer Fabrication Facility

Introduction

The 300 mm wafer fabrication facilities for industry were a precursor to today's advanced technology facilities. The evolution of semiconductor technology has affected the design of the cleanroom environment. These characteristics and requirements of a 300 mm wafer fabrication can be applied to similar issues in the design of buildings for advanced technology.

Wafer Fabrication Facilities

A 300 mm wafer fabrication plant is the latest generation of factories used by the semiconductor industry to manufacture microchips. The 300 mm designation refers to the diameter of the silicon wafer used for processing. 300 mm is approximately 12 in.

Of some 1000 operating fabrication plants (fabs) around the world, a dozen or so are currently producing on 300 mm wafers. Because of the relative newness and small number of such facilities, operating data is also scarce.

300 mm wafer fabs are truly huge factories. Typically, the cleanroom itself is 9300–14,000 m² (100,000–150,000 ft.²). The total facility is around 93,000 m² (1,000,000 ft.²) accounting for support areas, space for utilities and office areas. As an example, the fabrication building of one facility covers five football fields. This is the production cleanroom building only—it does not include the support areas, utility buildings, or offices.

Fig. 11.2 Relative cost per
chip, 200 mm vs. 300 mm
(courtesy HDR
Architecture, Inc.)

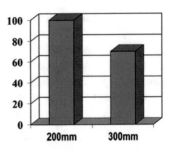

These facilities are high yield factories designed for maximum efficiency with up to 40,000 wafer starts per month. With as many as 500 process steps on each wafer, the opportunities for product contamination are enormous.

Not only are these facilities big, they are also expensive. A new fab today costs ~ $3 billion. Eighty to eighty-five percent of this cost is for production tools and the balance is the building itself. In order to be competitive in the marketplace, these costs have to be recovered in about a 3-year timeframe.

The design and construction schedule challenges are tremendous. In 1995, for a 200 mm fab, the time from start of design to first wafer was an average of 23 months. This included design, construction, tool installation, start-up, and process qualification. Meeting these goals took some innovative design and construction techniques. Jump forward 8 years to a 300 mm fab being built in 2003, that 23 months became only 15 months. This was extremely challenging for design and construction professionals, but if the schedules weren't met, the economic model for the factory didn't work.

The question becomes, if they're so expensive and challenging to build, why do it? A 300 mm fab can turn over $2 billion in annual revenues. The 300 mm wafer is 2.25 larger in area and squeezes approximately 2.5 times as many chips on each wafer when compared to a 200 mm wafer (see Fig. 11.2).

The cost per wafer is higher for the 300 mm factory but, because of the number of chips per wafer, the cost per chip is substantially less.

Today's microelectronics have in fact crossed the threshold into the nano realm. At the 90 nm technology node as defined by the 2001 edition of the *International Technology Roadmap for Semiconductors*, the 65 nm technology node was proven in the lab. In 2002, IBM announced a 6 nm silicon transistor. 300 mm factories are the first to transition to the nanoscale. It is not the wafer size so much as the technology being applied to this generation of semiconductor with which we are interested. And most importantly what this technology means to the cleanroom environment.

What Types of Contamination Are of Concern?

In the broad sense, contamination is any property within the wafer environment that will impact manufacturing yield such as:

Table 11.3 Typical cleanroom specs for a 300 mm fab

Filter ceiling	100 % coverage w/gel seal grid
Airflow	60–90 fpm (~0.3–0.45 m/s)
Temperature	72 °F ± 0.5 °F critical areas
	72 °F ± 2 °F non-critical areas
Humidity	50 % ± 2 % RH critical areas
Pressurization	0.05″ WG with respect to corridor

RH relative humidity, *WG* water gauge

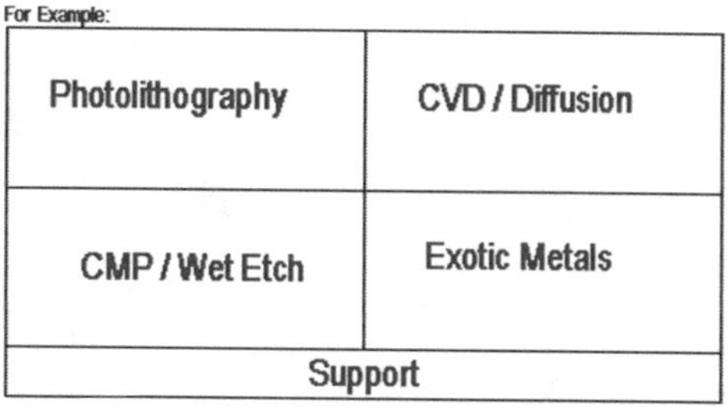

Fig. 11.3 Illustrative design (courtesy HDR Architecture, Inc.)

For Example:

Photolithography	CVD / Diffusion
CMP / Wet Etch	Exotic Metals
Support	

- Temperature
- Humidity fluctuations
- Vibration
- Static discharges
- Electromagnetic interference
- Radio frequency interference

Typical cleanroom specs for a 300 mm fab are given in Table 11.3.

A technique used frequently today is to provide segregated HVAC systems for different process areas to prevent cross-contamination. One 300 mm fab's design had a scheme as shown in Fig. 11.3.

A trend that changed everything was wafer isolation technology. The environmental requirements for 300 mm wafers are in the Class 0.1 range or ISO Class 2. This is essential if yields are to remain profitable. The problem is that it is very difficult to achieve an operational ISO Class 2 cleanroom throughout an entire fab. It is simply impractical to expose the wafer to the cleanroom environment regardless of class and expect zero contamination. The solution is to isolate the wafer from the cleanroom, and from the people, and from all the other sources of contamination (see Fig. 11.4).

The industry acknowledged that wafer isolation was an enabling technology below the 90 nm technology node. The wafer is never exposed to the cleanroom. It is sealed in a pod between process steps. The pod is known as the FOUP (Front Opening Unified Pod). Prior FOUPS contained cleanroom air but present models offer an inert gas purge.

The wafers remain in the FOUP until it is mated to the production tool. This interface happens at the load port. The next step in the wafer isolation is to isolate the tool within its own mini-cleanroom environment. It is much easier to maintain

Fig. 11.4 Cleanroom
contamination evolution
(courtesy HDR
Architecture, Inc.)

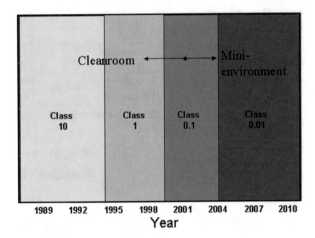

Fig. 11.5 Illustration of
FOUP layout (courtesy
HDR Architecture, Inc.)

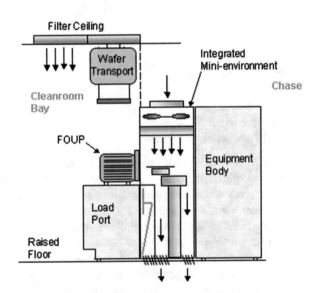

specification in a small contained space. Fan filter units above the tool provide filtered air stream and temperature/humidity control.

The final component to wafer isolation is the wafer handling system. This system automatically transports the FOUPS containing the wafers between the various production tools. Figure 11.5 displays the process schematically.

The wafer isolation technology has helped address issues with:

- Particulate contamination
- Temperature control
- Airborne Molecular Contamination

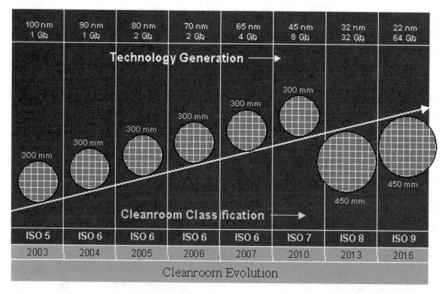

100 nm 1 Gb	90 nm 1 Gb	80 nm 2 Gb	70 nm 2 Gb	65 nm 4 Gb	45 nm 8 Gb	32 nm 32 Gb	22 nm 64 Gb
ISO 5	ISO 6	ISO 6	ISO 6	ISO 6	ISO 7	ISO 8	ISO 9
2003	2004	2005	2006	2007	2010	2013	2016

Fig. 11.6 International Technology Roadmap for Semiconductors (ITRS) roadmap for node generations (courtesy of HDR Architecture, Inc.)

The future cleanroom classification is decreasing according to the *International Technology Roadmap for Semiconductors*. The cleanroom is getting less clean. Many of today's 300 mm cleanrooms are, in fact, being designed to ISO 5/Class 100 standards. The roadmap calls for ISO 7/Class 10,000 cleanroom by 2010 (see Fig. 11.6).

The 300 mm wafer fab relies heavily on wafer isolation and reduced requirements on:

- Airflow
- Capital cost of the building
- Gowning requirement
- Operating costs for cleanroom

The continuing problem is that tools for 300 mm production are very large and generate a lot of heat. This, of course, runs counter to the requirement for tighter temperature controls. Designers are seeing that while the cleanroom classification can be reduced, we still need significant air quantities to remove heat gain. As an alternative, 300 mm tools are being designed to remove a larger percentage of heat with process cooling water.

In summary, as the technology advances, wafer isolation technology will allow the trade of an expensive, complex cleanroom for many expensive, complex, isolated production tools. This trades a field-assembled clean environment for a factory assembled clean environment with all its associated Quality Control measures. The *International Technology Roadmap for Semiconductors* has identified

450 mm as the next probable wafer size (see Fig. 11.6). The Roadmap also shows the required cleanroom environment dropping to ISO 9, which is the class that was added above Class 100,000.

The ultimate goal may be the "lights out" fab where all processing is done robotically without the need for humans. While the fight against contamination continues, there are other changes going on in the 300 mm cleanroom environment. In addition to tighter control of particulate contamination, there are other trends at which to look. Tighter temperature controls are required due to the fine geometries. A very small temperature change can have a detrimental effect on mask alignment, thereby ruining proper fabrication of a wafer. Temperature specification is down to ± 0.5 °F in the cleanroom. This is not as tight as some of the measurement lab criteria, but over a large area, it is a challenge. Within the lithography equipment, temperature control of ± 0.1 °F is required. Process cooling water is being used more in 300 mm tools to help remove the large heat loads generated by the equipment. Filtration of atmospheric molecular contamination (AMC) is now a requirement, not an option. Levels of amines in lithography equipment are in the 1–10 ppb range. Because of the high volume of air recirculated in today's massive cleanrooms, high-efficiency fans and motors are commonplace to reduce energy costs.

Part B: Building Statistics[6]

Introduction

For completeness, the editors thought that some overall building statistics would be useful for readers. Toward this end the statistics on building area, building costs, and building performance characteristics are given for the following buildings:

- Birck Nanotechnology Center—Purdue University, West Lafayette, Indiana
- National Institute for Nanotechnology—University of Alberta, Edmonton, Canada
- Nanoscience Building—Naval Research Laboratory, Washington DC
- Center for Integrated Nanotechnologies—Sandia National Laboratories, Albuquerque, New Mexico, and Los Alamos National Laboratory, Los Alamos, New Mexico
- Molecular Foundry—Lawrence Berkeley National Laboratory (LBNL), Berkeley, California
- Advanced Measurement Laboratory—National Institute of Standards and Technology, Gaithersburg, Maryland

---

[6] Specifications given for these facilities are those available at the time of drafting the text. Present building performance specifications will be different.

Birck Nanotechnology Center

Purdue University, West Lafayette, IN

Area	19,986 m^2
Project budget	$57 million
Architect/engineer	HDR Architecture, Inc.

Birck Nanotechnology Research Center is the first of its kind in the state of Indiana. The facility provides approximately 20,000 gross square meters of interactive, interdisciplinary laboratory, cleanroom, office, teaching laboratories and seminar space to pursue research. It anchors Purdue's new Discovery Park located on the southwest edge of the West Lafayette campus. Programs involve undergraduate teaching, graduate research and technology transfer initiatives with industry partners. The Schools of Biology, Chemistry, Physics, and other engineering disciplines participate in research efforts (Fig. 11.7).

Building areas	
Level 1	8535 m^2
Level 2	6675 m^2
Subtotals	15,210 m^2
Mechanical penthouse	4475 m^2
Total	19,986 m^2

Environmental criteria	
General laboratory statistics	
Vibration	50–100 μm/s
Noise criteria	NC 35–NC 45
Temperature	21±1 °C
Relative humidity	30–50 %
Cleanroom class 10 statistics	
Vibration	3 μm/s
Noise criteria	NC 55–NC 65
Temperature	20±1 °C
Relative humidity	45±5 %
Cleanroom class 100 statistics	
Vibration	6 μm/s
Noise criteria	NC 55–NC 65
Temperature	20±1 °C
Relative humidity	45±5 %
Cleanroom class 1000 statistics	
Vibration	25 μm/s
Noise criteria	NC 55–NC 65
Temperature	20±1 °C
Relative humidity	45±5 %

(continued)

Environmental criteria	
Nanostructures lab I–III	
Vibration	3 μm/s
Noise criteria	NC 30
Temperature	21±0.1 °C
Relative humidity	45±5 %

Fig. 11.7 Birck Nanotechnology Research Center (courtesy of HDR Architecture, Inc.)

National Institute for Nanotechnology

University of Alberta, Edmonton, Canada

Area	24,353 m^2
Project budget	$40 million
Architect/engineer	The Cohos Evamy Partners (HDR Nanotech Consultants)

The National Institute for Nanotechnology (NINT)—a collaborative venture of the National Research Council (NRC), the Province of Alberta, and the University of Alberta—will be a national center for nanotechnology research in Canada. Functions include 10,500 m^2 of generic laboratories, cleanroom and metrology suites. The metrology suites are vibration-isolated, temperature-controlled spaces (Fig. 11.8).

Building areas	
Gross area	24,353 m^2
Net area: NINT	10,725 m^2
Univ. of Alberta	3600 m^2
Net/Gross	58.8 %

Environmental criteria	
Mechanical vibration	<3 μm/s
Electromagnetic interference	<0.1 mG rms
Temperature variations	<0.1 °C/h
Air-pressure fluctuations	<few Pascal/min
Air flow across instrument	<10 m/min
Clean Rooms	Class 100–Class 10,000

65 % is generic laboratory and support space
22 % is cleanroom and high-end characterization space

Fig. 11.8 NINT Facility at Univ. of Alberta (courtesy of NINT)

Naval Research Laboratory Nanoscience Building

NRL, Washington, DC

Area	1394 m^2
Project budget	$15 million
Architect/engineer	Gannett Fleming

The Naval Research Laboratory's Nanotechnology Building is the first of its kind for the Department of Defense. The facility provides approximately 1394 m^2 of interactive, interdisciplinary measurement laboratory and cleanroom research space. The 465 m^2 cleanroom (with three interconnected compartments) is located on the center axis with all air handling overhead; the measurement laboratories (each on its own individual slab) are separated/isolated from the cleanroom by 4 m wide utility corridors designed to house noisy equipment. Office and meeting space was deliberately excluded from the building to reduce sources of noise. External walls are windowless and insulated to buffer surrounding noise sources (Fig. 11.9).

Building areas	
Single floor operational space (slab on grade)	
Class 100 general purpose cleanroom	279 m^2
Class 100 lithography (charcoal filter)	93 m^2
Class 100 patterning	93 m^2
Quiet module (8 @ 37 m^2)	297 m^2
Ultra-quiet Module (4 @ 37 m^2)	167 m^2

Environmental criteria	
Cleanroom	
Vibration	Unspecified
Acoustic noise	NC55–NC65
Temperature	20 °C±0.5 °C
	20 °C±0.1 °C Patterning compartment
Relative humidity	45±5 %
Quiet modules	
Vibration	NIST type A
Acoustic noise	<NC35
EMI	<0.3 mG pp @60 Hz
Temperature	20 °C±0.5 °C
Relative humidity	45±5 %
Ultra-quiet modules—measurement compartment	
Vibration	NIST type A
Acoustic noise	<NC 25
EMI	<0.3 mG pp @60 Hz
Temperature	20 °C±0.1 °C
Relative humidity	45±5 %

All building air handling apparatus located over cleanroom on separate pilings 4 m wide utility corridors adjacent to modules for location of noisy equipment

Restricted building access via corridor with interlocked double doors to reduce pressure fluctuations

The ultra-quiet modules have a control compartment of 14 m^2 separated from a measurement compartment of 23 m^2 which has the most stringent noise specifications

Fig. 11.9 NRL Nanoscience Building (courtesy of NRL)

Center for Integrated Nanotechnologies

Sandia National Laboratories, Albuquerque, NM; Los Alamos National Laboratory, Los Alamos, NM

Area	12,317 m^2
Project budget	$76 million
Architect/engineer	HDR Architecture, Inc.

The Center for Integrated Nanotechnologies (CINT) is a Department of Energy/ Office of Science Nanoscale Science Research Center jointly operated as a national user facility by Sandia National Laboratories (SNL) and Los Alamos National Laboratory (LANL) (Fig. 11.10).

Its focus is on establishing the scientific principles that govern the design, performance, and integration of nanoscale materials. Through its Core Facility and Gateways to both Los Alamos and Sandia National Laboratories, CINT provides access to tools and expertise to explore the continuum from scientific discovery to the integration of nanostructures into the micro and macro worlds.

One of the primary goals of CINT is to provide an interdisciplinary environment that can serve the scientific community, including those from the government, university, and private sectors.

Building areas (core facility)	
Synthesis lab wing	837 m^2
Characterization lab wing	837 m^2
Cleanroom	837 m^2
Office/conf rooms/circulation	3997 m^2
Mechanical/general	2417 m^2
Total	8924 m^2

Environmental criteria (core facility)	
Typical laboratory statistics	
Area	1673 m^2
Ceiling height	3.05 m
Air classification	Unspecified
Vibration	125 µm/s
Temperature	70 °F±2 °F
Relative humidity	40 %±10 %
Cleanroom statistics	
Area	837 m^2
Air classification	1000
Ceiling height	3.05 m
Vibration	125 µm/s
Temperature	70 °F±0.5 °F
Relative humidity	40 %±10 %

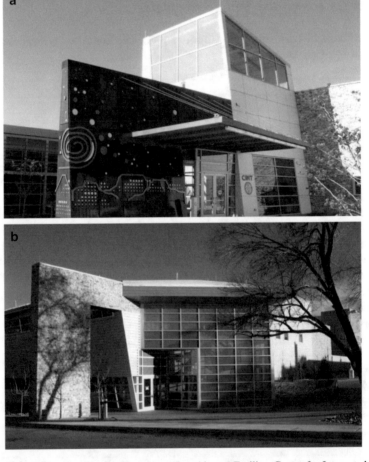

Fig. 11.10 (**a**) Core Facility. (**b**) Gateway to Los Alamos Facility, Center for Integrated Nano-technology Facilities (courtesy of HDR Architecture, Inc.)

Molecular Foundry

Lawrence Berkeley National Laboratory, Berkeley, CA

Area	8738 m^2
Project budget	$52 million
Architect/engineer	The Smith Group

Organized into six interdependent research facilities, the Molecular Foundry (http://foundry.lbl.gov/six/affiliated.html) and its affiliated research laboratories provide access to state-of-the-art instrumentation, scientific expertise and specialized techniques to help users address the myriad challenges in nanoscience and nanotechnology (Fig. 11.11).

Building areas
The research facilities are housed in a six-story, 8738 m^2 building equipped with advanced, sometimes one-of-a-kind instruments. It includes approximately 446 m^2 of Class 100 cleanroom space, with a smaller Class 10 area for nanofabrication/lithography and clean measurement, and a 511 m^2 low vibration, low-electromagnetic-field laboratory housing state-of-the-art imaging and manipulation tools

Environmental criteria	
Basement lower level (LL2)	
Area	1180 m^2
Ceiling height	6 m
Vibration	50 μm/s
Temperature	21 °C±1 °C
Relative humidity	30–60 %
Basement lower level (LL1)	
Area	1600 m^2
Ceiling height	6 m
Vibration	3–50 μm/s
Temperature	20 °C±1 °C
Relative humidity	30–60 %
Level 1	
Area	1400 m^2
Ceiling height	4.9 m
Vibration	50 μm/s
Temperature	22 °C±1 °C
Relative humidity	30–60 %
Levels 2–4	
Area	1350 m^2
Ceiling height	4.9 m
Vibration	3–50 μm/s
Temperature	22 °C±1 °C
Relative humidity	30–60 %

Fig. 11.11   Molecular foundry (courtesy of Lawrence Berkeley National Laboratory)

Advanced Measurement Laboratory

National Institute of Standards and Technology, Gaithersburg, MD

Area	47,480 m^2
Project budget	$235 million
Architect/engineer	HDR Architecture, Inc.

The most environmentally stable facilities in the world can now be found at the new Advanced Measurement Laboratory (AML) of the National Institute of Standards and Technology (NIST). The facility gives NIST and its partners in U.S. industry access to research and development capabilities unequalled in the world for research in such twenty-first century applications as nanotechnology, semiconductors, biotechnology, advanced materials, quantum computing, and advanced manufacturing. Research at NIST covers a broad spectrum. Over 100 different types of laboratories were identified during initial AML user interviews. Classifying the spaces into a few general categories helped to establish a modular concept for lab space organization and utility distribution. With this modular approach, space and utilities can be readily modified for changing research efforts of the future. The building, completed in December 2003 and officially opened in June 2004, features five separate wings: Two metrology labs below grade, two instrumental labs and a connecting cleanrooms building. The design accounts for stringent environmental controls on particulate matter, temperature, humidity, vibration and electromagnetic interferences. Since this facility provides examples of the more demanding requirements, more details are provided in the following (Fig. 11.12).

Building areas	
Instrument East	9529 m^2
Instrument West	11,858 m^2
Metrology East	8470 m^2
Metrology West	9103 m^2
Cleanroom	8520 m^2

Environmental criteria	
Temperature control	±0.25 °C base control
Laboratory control	±0.1 °C or ±0.01 °C
Humidity control	±1 to 5 %
Air quality	Class 100–1000
Vibration control	0.2–3 μm/s
Class 100, upgradeable to Class 10	
Total AML	47,480 m^2
Net/Gross ratio	41.6 %

48 precision temperature control labs
18 extremely low vibration labs

Fig. 11.12 AML (courtesy of NIST)

National Institute of Standards and Technology, Advanced Measurement Laboratory

Facts and Figures

Architectural

- 151 metrology lab modules; 187 instrument lab modules

Class 100 Cleanroom Facility

- All laboratories meet baseline vibration criterion of "NIST-A". 18 Metrology labs meet vibration criterion of "NIST-A1" (Baseline velocity amplitude of 3 µm/s, down to 0.75, 0.5 µm/s or less). Built-in provisions for future "NIST-A1" slabs in ten additional laboratories.
- Acoustical criterion NC-30–NC-45
- Service galleys between laboratory modules provide maximum flexibility
- 51 chemical fume hoods, 6 clean laminar flow hoods, and 10 biological safety cabinets

Structural

- Labs isolated from adjoining labs and other spaces for vibration control
- Accessible walk-on ceiling above the labs for easy maintenance
- Removable steel tube ceiling support structure allowing vertical expansion of the laboratory space from a baseline height of 4 m to over 7 m
- A column-free cleanroom space with a rigid, elevated waffle slab that simulates slab on grade conditions
- Multiple layers of vibration isolation in the structural design
- Special concrete isolation slabs supported on air springs with active or passive controls

Electrical

- Five sets of three transformer spot networks
- A dedicated switchboard to lab equipment
- Two panels per lab served by dedicated transformer
- Two isolated redundant UPS systems
- Transient voltage surge suppression at each lab panel and main switchboard
- 1000 kW emergency generator with double spring vibration isolation
- Extremely low frequency electromagnetic interference mitigation

Mechanical

- Baseline temperature control in laboratories of ± 0.25 °C
- Special temperature control to ± 0.1 °C or ± 0.01 °C for 48 high-accuracy laboratories
- Low temperature chillers for humidity control
- Chiller heat recovery system to minimize cost of dehumidification
- Clean steam generators for humidification
- 116 industrial-grade air-handling units circulating 70,000 m^3/min (2,500,000 CFM)
- Cleanroom with 26 circulating air-handling units
- High-efficiency filtration with 95–99.97 % efficiency

Plumbing

- Separate water and drainage systems
- Reagent-grade pure water to labs and cleanroom
- Laboratory vacuum system and compressed air system
- Safety showers and eye washes with dedicated tempered water
- Natural gas and gaseous nitrogen distribution system to labs
- Gaseous high-purity and liquid nitrogen to cleanroom
- Special gases cabinets and piping in cleanroom

Fire Protection

- Complete fire detection and fire alarm system
- Toxic gas bunker area with blowout panels
- Complete toxic gas monitoring system in cleanroom

Acknowledgement Thanks to Shannon Jones, Larry Allard, Mark Jamison, and Tom Isabell, who provided information on the NINT, ORNL, and 300 mm and analytic facilities, respectively, that was used for the initial draft of those sections.

Appendix A. Workshop Agendas

Buildings for Advanced Technology I
January 14–16, 2003
National Institute of Standards and Technology
Gaithersburg, Maryland

Tuesday January 14, 2003	
11:00 am–1:00 pm	REGISTRATION
1:00–1:40 pm	INTRODUCTION SESSION
	Introduction—Clayton Teague, NIST
	Buildings for Advanced Technology—Ahmad Soueid, HDR Architecture, Inc.
1:40–3:30 pm	ACOUSTIC ISOLATION AND CONTROL OF PRESSURE FLUCTUATION SESSION
	Moderator: Hal Amick, PE, Colin Gordon and Associates
	Building Design for Advanced Technology Instruments Sensitive to Acoustical Noise—Michael Gendreau, Colin Gordon and Associates
	Environmental Enclosure for the MIT Nanoruler—Mark Schattenburg, MIT
	Acoustic and Vibration Control in Vacuum: A Case Study—John Lawall, NIST
	Debugging Acoustical Interference; Is it from the Building or the Tool? A Case Study—Ron Reifenberger, Purdue University
3:30–3:45 pm	Break

(continued)

© Springer International Publishing Switzerland 2015
A. Soueid et al. (eds.), *Buildings for Advanced Technology*, Science Policy Reports,
DOI 10.1007/978-3-319-24892-9

3:45–5:45 pm	VIBRATION ISOLATION SESSION
	Moderator: E. Clayton Teague, NIST
	Vibration Isolation at Building Level—Eric Ungar, PE, Acentech, Incorporated
	Isolating Instruments from Building Vibration—Hal Amick, PE, Colin Gordon and Associates
	Latest Vibration Isolation Techniques at Instrument Level Quiet Tables and New Technologies—Rod Horning, TMC Manufacturing
	Design and Operation of the Nanoscale Physics Facility in the NIST Physics Laboratory—Joe Stroscio, NIST
5:45–7:00 pm	Reception

Wednesday January 15, 2003

8:00–10:00 am	TEMPERATURE AND HUMIDITY CONTROL SESSION
	Moderator: James Murday, National Nanotechnology Initiative, Naval Research Laboratory
	HVAC Design for Nanotechnology Buildings—Ted Zsirai, PE, HDR Architecture, Inc.
	Measurements of Temperature Stability and Uniformity in Several Types of Laboratories—Bea Sennewald, HDR Architecture Inc. and Julian Hunt, National Physical Laboratory
	Case Studies of Precision Temperature Control Systems for Air Showers and Liquids at Lawrence Livermore National Laboratory—Jeff Roblee, Precitech, Inc.
	NIST Temperature Controlled Laboratory Module—Steve Treado, NIST
10:00–10:15 am	Morning Break
10:15 am–12:15 pm	AIR CLEANLINESS SESSION
	Moderator: James Whetstone, NIST
	300 mm Wafer Fab Contamination Control—Mark Jamison, HDR Architecture, Inc.
	Organic and Chemical Contamination in Advanced Laboratories—Michael Somin, Earl Walls and Associates
	Achieving ISO Class 3 in a Retrofit Cleanroom: A Case Study—John Weaver, Delphi Delco Electronics Systems
	Matrix Development: A Prerequisite for a Successful Cleanroom Design—Tim Loughran, AdvanceTEC, LLC

(continued)

12:15–1:15 pm	Lunch
1:15–3:15 pm	ELECTRO MAGNETIC INTERFERENCE/RADIO FREQUENCY INTERFERENCE SESSION
	Moderator: Michael Gendreau, Colin Gordon & Associates
	Interactions between Nanomeasurements and Nanobuilding Design—Bob Erdman, Erdman Measurement Consulting
	EMI/RFI: Cause; Site Analysis; Building Evaluations; Mitigation Solutions—Lou Vitale, Vitatech Engineering
	EMI Issues in a University Microelectronics/Nanotechnology Laboratory—Tim Miller, Purdue University
	Scanning Electron Microscopy in Real-World Environments—Andras Vladar, NIST
3:15–3:30 pm	Afternoon Break
3:30–4:30 pm	TECHNICAL SURVEY SESSION
	Moderator: E. Clayton Teague, NIST
	Technical Survey—Clayton Teague, NIST

Thursday January 16, 2003

8:00–10:25 am	POWER CONDITIONING AND GROUNDING
	Moderator: Michael Gendreau, Colin Gordon & Associates
	Designing for Clean Power—Dave Bechtol, HDR Architecture, Inc.
	Cost-Effective Power Conditioning for Advanced Technology Buildings—Mark Stephens, EPRI PEAC Corporation
	Grounding Needs of Instrumentation—Bob Erdman, Erdman Measurement Consulting
	Why all the Noise about Grounding?—Ralph Morrison
10:25–10:40 am	MORNING BREAK
10:40 am–12:40 pm	SYSTEMS INTEGRATION
	Moderator: Ahmad Soueid, HDR Architecture, Inc.
	Science Needs for High Performance Laboratories—Kamal Hossein, NPL
	The Aberration Corrected Electron Microscope (ACEM) at ORNL—Larry Allard, Oak Ridge National Laboratory
	Constructing Advanced Technology Buildings—James V. Bartlett, Jr. PE, NIST and Todd Snouffer, NIST
	Systems Integration and Competing Criteria—Ahmad Soueid, HDR Architecture, Inc., Dave Bechtol, HDR Architecture, Inc. and Hal Amick, Colin Gordon and Associates
12:40–1:30 pm	BREAK
1:30–3:30 pm	NIST ADVANCED MEASUREMENT LABORATORY CONSTRUCTION SITE TOUR
	NIST Advanced Measurement Laboratory Overview—Ahmad Soueid, HDR Architecture, Inc.
	AML Construction Site Tour No. 1
	AML Construction Site Tour No. 2

Buildings for Advanced Technology II
January 21–23, 2004
Mesa, Arizona

Wednesday January 21, 2004	
Defining Science Needs	
M/Cs: Ahmad Soueid and Dr. Allan Chasey	
10:00 am–noon	Registration
11:00 am–noon	Welcome Lunch Buffet
	Workshop Introduction—Dr. Greg Raupp, Vice President, Office of Research and Economic Affairs, Arizona State University
12:15–1:00 pm	Keynote Speaker: Overview of the National Nanotechnology Initiative, Nanotechnology Research Directions—Dr. E. Clayton Teague, Director, National Nanotechnology Coordination Office
	Day 1 Introduction—Dr. Peter Crouch, Dean, Ira A. Fulton School of Engineering, Arizona State University
1:00–2:00 pm	Brookhaven National Laboratory, Center for Functional Nanomaterials—Dr. Robert Hwang, Director of Center for Functional Nanomaterials, Brookhaven National Lab
2:00–3:00 pm	Sandia National Laboratories, Center for Integrated Nanotechnologies (CINT), A New Model for a Nanoscience Research User Facility—Dr. Neal D. Shinn, Manager of the Surface and Interface Science Department, Sandia National Laboratories
3:00–3:15 pm	Break Refreshments, sponsored by Mc Carthy and ASU Technopolis
3:15–4:15 pm	Research facilities to serve the needs Of twenty-first century Biology: A Nexus Of Disciplines, Technologies and Discoveries—Dr. Michael Knotek, Consultant, Knotek Consulting
4:15–5:15 pm	Interdisciplinary Innovation At The Nano-Scale And Its Impact On Laboratory Construction—Dr. Frederic Zenhausern, Arizona State University
5: 15–6:45 pm	Reception Networking Reception sponsored by HDR Architecture, Inc. and CREATE

Thursday January 22, 2004	
Outlining Trends/Developing Solutions	
M/C: Ahmad Soueid	
7:30–8:00 am	Breakfast Continental, sponsored by Mc Carthy and M+W Zander
8:00–8:15 am	Day 2 Introduction—Ahmad Soueid, Principal/Senior Vice President HDR Architecture, Inc.
8:15–9:00 am	Analytical Instrumentation Facility Requirements For Nanotechnology—Dr. Tom Isabell, Assistant TEM Product Manager, JEOL USA, Inc.
9:00–10:15 am	Architectural Trends, Solutions, and Sustainable Design—Ahmad Soueid, Moderator
	Tom Gerbo, Vice President, HDR Architecture, Inc.
	Ken Filar, Senior Architect, M+W Zander
	David Gibney, Sustainable Design Coordinator, HDR Architecture, Inc.

(continued)

10:15–10:30 am	Break Refreshments, sponsored by Colin Gordon & Associates
10:30–11:45 am	Mechanical, Process Engineering Solutions—Steve Riojas, Moderator
	Norm Toussaint, Senior Mechanical Engineer, HDR Architecture, Inc.
	Chris Case, Project Manager, Affiliated Engineers, Inc.
	William Acorn, Principal and Founder, Acorn Consulting Services
11:45 am–1:00 pm	Lunch Buffet, sponsored by HDR Architecture, Inc. and CREATE
1:00–2:00 pm	Electrical, EMI, Grounding—Dr. David Janes, Moderator
	Dave Bechtol, Electrical Section Manager, HDR Architecture, Inc.
	Lou Vitale, President and Chief Engineer, Vitatech Engineering, LLC.
2:00–3:00 pm	Mechanical Systems Noise Issues: Case Studies—Steve Westfall, Moderator
	Hal Amick, VP, Technology Development, Colin Gordon and Associates
	Amir Yazdanniyaz, Principle Acoustics Consultant, Arup Acoustics
3:00–3:15 pm	Break Refreshments, sponsored by Colin Gordon & Associates and Abbie Gregg, Inc.
3:15–5:00 pm	Meeting User Requirements: Case Studies—Tom Gerbo, Moderator
	Dr. David Janes, Associate Professor of Electrical and Computer Engineering, Purdue University
	Phil Haswell, Director of Facilities, University of Alberta, Canada
	Jack Stellern, Senior Project Manager, Oak Ridge National Laboratory

Friday January 23, 2004

Protecting the Investment

M/C: Dr. Allan Chasey

7:30–8:00 am	Breakfast Continental
	Day 3 Introduction—Dr. William Badger, Director, Del E. Webb School of Construction, Arizona State University
8:00–9:15 am	Scope Development, Programming, Cost, Cost Control, Project Control
	Abbie Gregg, AGI
	Greg Parker, Currie and Brown
9:15–9:30 am	Break Refreshments, sponsored by Abbie Gregg, Inc. and M+W Zander
9:30–10:15 am	Project Management (contract delivery, commissioning, quality, expertise)
	Robert Harper, Gilbane Building Company
	Michael Christeson, Gilbane Building Company
10:15–11:00 am	ASU AZ Bio Design Facility
	Brett Dominguez, DPR
	Brett Helm, DPR
	Terry Abair, Sundt Construction, Inc.
11:30 am–noon	Wrap Up

Buildings for Advanced Technology Workshop III
February 6–8, 2006
Purdue University
West Lafayette, Indiana

Monday February 6, 2006

2:00–3:00 pm	Registration
	Technical Panel—NNI at 5 Years: The Transformation
3:00–3:15 pm	NNI Vision and Outcomes—Mihail C. Roco
3:15–3:30 pm	Progress in Modelling and Simulation at the Nanoscale—Peter T. Cummings
3:30–3:45 pm	NNI Impact on Industry—George Scalise
4:15–4:30 pm	Venture Capital Investments in Nanotechnology—Josh Wolfe
4:30–4:45 pm	Nanofabrication—R. Fabian W. Pease
4:45–5:00 pm	Engineering in Nanoelectronics—Phaedon Avouris
	Lecture Forum—NNI at 5 Years: Public Interest
5:30–6:00 pm	Welcome—Martin C. Jischke
6:00–6:15 pm	Nanotechnology for Cancer Treatment—Gregory J. Downing
6:15–6:30 pm	Societal Implications of Nanotechnology—David Guston
6:30–6:45 pm	Public Engagement and Societal Implications—Vivian Weil
7:00–8:30 pm	BAT III Workshop Reception and Dinner

Tuesday February 7, 2006

8:00–8:25 am	Workshop Introduction/Overview—Clayton Teague and Ahmad Soueid
8:25–9:00 am	Keynote—Sally Mason, Provost, Purdue University
9:00–9:30 am	Purdue University: Overview of Purdue's Discovery Park and Birck Nanotechnology Center—Jim Cooper and George Adams
9:30–10:00 am	Purdue University: Birck Nanotechnology Center's Laboratories: Defining User Requirements—David Janes and Ron Reifenberger
10:00–10:30 am	Morning Break, Stewart Center West Lobby
10:30–11:00 am	Purdue University: Birck Nanotechnology Center's Biological Laboratories and Nanofabrication Cleanrooms—Rashid Bashir and John Weaver
11:00–11:30 am	National Institute of Standards and Technology: NIST Advanced Measurement Laboratory—Todd Snouffer
11:30 am– 12:00 pm	University of California, Riverside: UCR Materials Science and Engineering Building (MSE) Project—Luis Carrazanna
12:00–1:30 pm	Lunch
1:30–2:00 pm	CEA—MINATEC (France): Centre of Innovation in Micro and Nanotechnology—Jean-Charles Guibert
2:00–2:30 pm	University of Alberta/National Research Council (Canada): National Institute of Nanotechnology (NINT)—Phil Haswell
2:30–3:00 pm	National Nanofab Center (Korea): NNFC Design and Construction—Hee Chul Lee
3:00–3:30 pm	University of Waterloo (Canada): Quantum Nano Centre—Ray Laflamme
3:30–4:00 pm	Afternoon Break, Stewart Center West Lobby

(continued)

4:00–4:30 pm	Panel—Purdue University Measured Vibration and EMI Technical Performance of Birck Center Laboratories and Cleanroom.
	Moderated by: David Janes: (Reifenberger, Soueid, Vitale)
4:30–5:00 pm	Panel—Purdue University Planning, Design, Construction and Commissioning of the Birck Nanotechnology Center
	Moderated by: Luh M. Chang. (Janes, Hatke, Wallace, Skiba, Weaver)
5:00–5:15 pm	Adjourn
5:15–7:15 pm	Tour—Purdue University Birck Nanotechnology Center

Wednesday February 8, 2006	
8:00–8:15 am	Day 2 Overview—Ahmad Soueid
8:15–9:00 am	Keynote: The U.S. National Nanotechnology Initiative—Clayton Teague
9:00–9:30 am	Naval Research Laboratory: Naval Research Laboratory Nanoscience Building—Jim Murday
9:30–10:00 am	U.S. Department of Energy: DOE Nanoscale Science Research Centers—Linda Horton
10:00–10:30 am	Morning Break, Stewart Center West Lobby
10:30–11:00 am	Oak Ridge National Laboratory: ORNL Center for Nanophase Materials Sciences (CNMS)—Linda Horton
11:00–11:30 am	Sandia National Laboratories: SNL Center for Integrated Nanotechnologies (CINT)—Neal Shinn
11:30 am–12:00 pm	Brookhaven National Laboratory: BNL Center for Functional Nanomaterials (CFN)—Mike Schaeffer
12:00–1:30 pm	Lunch
1:30–2:00 pm	Argonne National Laboratory: ANL Center for Nanoscale Materials—Derrick Mancini
2:00–2:30 pm	Oak Ridge National Laboratory: ORNL Advanced Microscopy Project—Larry Allard
2:30–3:00 pm	Argonne National Laboratory: Vibration and Temperature Stability at the ANL Center for Nanoscale Materials (CNM) and ANL Advanced Photon Source (APS)—Marvin Kirshenbaum and John Sidarous
3:00–3:30 pm	Top Five Individual Working Group Sessions
3:30–4:00 pm	Afternoon Break, Stewart Center West Lobby
4:00–5:00 pm	Town Hall Meeting/Panel Top Five List Discussion Panel/Open Forum
	Moderated by: Ahmad Soueid (Horton, Stellern, Kirshenbaum, Fallier, Dyling, Allard)

Appendix B. Participants and Contributors

Terry Abair, Sundt Construction, Inc.; II	James Cooper, Purdue University; III
William R. Acorn, Acorn Consulting Services; II	Robert Cornwall, Pennsylvania State University; I
George Adams, Birck Nanotechnology Center, Purdue University; III	Alan Craig, Montana State University; I
Larry Allard, Oak Ridge National Laboratory; I, III	Ken Crane, Purdue University Construction Dept.; I
William Anderko, University of Pennsylvania; III	John Crate, The Cohos Evamy Partners; I
Hal Amick Colin Gordon & Associates; I, II	Mike Crommie, University of California, Berkeley; I
Helen Anderson, University of Pennsylvania; II	Peter Crouch, Arizona State University College of Engineering and Applied Sciences; II
Chris Andrews, C.L. Andrews, Inc; II	Andras Cserhati; I
Gus Ardura, HDR Architecture, Inc; II	Robert Cunningham, University of Notre Dame; III
William Bader, Cornell University; I	Michael Daly, University of Notre Dame; III
Bill Badger, Arizona State University; II	Theda Daniels-Race, Duke University;
Helen Bailey, University of North Texas; I	Nick Davis, The Beck Group; II
Keith Bailey, HDR Architecture, Inc; III	Suzanne Davis, Pepper Construction Company of Indiana; III
Barry Barker, Lab. for Physical Sciences; I	William Deckert, IPS; I
James Bartlett, Bartlett Consulting; I	Kim Dengler, University of Pennsylvania; III
Rashid Bashir, Purdue University; III	Leonard. Deptula, Bulley & Andrews, LLC; I, II
Ahmad Bayat, Vibro-Acoustic Consultants; II	Bernie Deutsch, Deutsch Associates; II
Phil Beadle, HDR Architecture, Inc.	Jeff Dibattista, The Cohos Evamy Partners; I
Dave Bechtol, HDR Architecture, Inc.; I, II, III	Lon Dill, Lab for Physical Sciences; I
Tina Benedetti, HDR Architecture, Inc; II	Christopher DiPado, IPS; I
Lisa Berkey, Penn State University; III	Brett Dominguez, DPR Construction, Inc.; II

(continued)

© Springer International Publishing Switzerland 2015
A. Soueid et al. (eds.), *Buildings for Advanced Technology*, Science Policy Reports,
DOI 10.1007/978-3-319-24892-9

Michael Bird, Brookwood Program Management; II, III	Keith Douglas, The Whiting-Turner Contracting Co.; III
Paul Blaum, GSFIC; III	F. Gordon Driedger, University of Alberta; I
Rick Black, Mallory + Evans, II	John Duconge, Georgia Institute of Technology; III
Don Bloomfield, HDR Architecture, Inc.; I	Patrick Dwyer, Rice University; I
Charles Brown, University of Illinois at Chicago; II	Ove Dyling, Brookhaven National
Dave Calcaterra, Deutsch Associates; II	Laboratory; I, II, III
Miguel Camacho, HDR Architecture, Inc.; II	Charles Edwards, The Beck Goup; II
Robert Campbell, Hemisphere Engineering, Inc.; I	Laura Ellington, Affiliated Engineers, Inc.; II
Craig Carey, Sundt Construction ,Inc.; II	Bob Erdman, Erdman Measurement Consulting; I
Luis Carrazana, University of California, Riverside; III	Tom Ewing, University of Pennsylvania—SAS; III
Chris Case, Affiliated Engineers, Inc.; II	Martin Fallier, Brookhaven National Laboratory; I, III
Dudley Caswell, Enterprise Innovations; I	Patrick Fay, University of Notre Dame; III
Louis Chan, Lawrence Berkeley National Laboratory; II	Ken Filar, M+W Zander; II
Luh Chang, Purdue University; I, III	Curt Finfrock, M+W Zander; I
Lee Chapman, IDC Architects; II	Richard Flanigan, Wafertech; III
Allan Chasey, Arizona State University; II	Dan Fontana, DL Withers Construction; II
Steven Cheung, Kinetic Systems, Inc; II	Queen-Mein Foo, HDR Architecture, Inc.; I
John Chisholm, Adolfson & Peterson Construction; II	David Fortin, University of Alberta; II
Mike Chlopek, The Whiting-Turner Contracting Co.; III	Daren Fornasero, PCL Construction Management; I
Michael Christeson, Gilbane Building Company; II	Robert Fuller, Hunt Construction Group, Inc.; III
Robert Chunko, FEI Company; I	Guillermo Gabrielli, Halsall Associates, Ltd.; III
Donna Clare, Cohos Evamy; I, II	Ed Galindo, CDM; III
Ken Clarke, PCL Construction Management; I	Ross Garcia, Los Alamos National Laboratory; I, II
Philip Cleary, Philip Morris; I	Jeff Gates, Intelligent Enclosures, Inc.; I
Joseph Cometa, FEI Company; I	Greg Gehlen, The Staubach Company; II
David S. Cooper, The Beck Group; II	Michael Gendreau, Colin Gordon & Associates; I
Tom Gerbo, Sears Gerbo Architecture; II, III	Charles Geraci, NIOSH; III
David Gibney, HDR Architecture, Inc; II	Marvin Kirschenbaum, Argonne National Laboratory; I, III
Anne Graham, Intel Corporation; II	Michael Knotek, Knotek Consulting; II
Trip Grant, Flad & Associates; III	Paul Knowles, PCL Construction Management; I
Abbie Gregg, Abbie Gregg, Inc.; II	John Kraft, HDR Architecture, Inc.; I

(continued)

Bud Guest, McCarthy Building Cos.; III	Jim Krupnick; II
Jean Guibert, CEA Minatec; III	Raymond Laflamme, Institute for Quantum Computing; III
Carl M. Hair, Hardin Construction Company; II	Michael Lam, National Research Council Canada; I
Mitchell Hall, KPMB Architects; III	John Lawall, NIST; I
Nancy Hansis, Flad & Associates; I	Barbara Lambing, HDR Architecture, Inc; II
Joseph Harkins, Lawrence Berkeley National Laboratory; I, II	Rayford Law, Kallmann McKinnell & Wood Architects; III
Jennifer Harms, HDR Architecture, Inc; II	DongWoo Lee, Human & Technologies Architects and Engineers; III
Bob Harper, Gilbane Building Company; II	Hee-Chul Lee, National NanoFab Center; III
Bob Harrington, Fluor Corporation; II	Tim Lee, Public Works & Govt. Services, Canada; I
Randy Harrison, DL Withers Construction; II	Dan Lehman, Pennsylvania State—Materials. Research. Institute; I
Phillip Haswell, University of Alberta; I, II, III	Rich Levey, Sundt Construction, Inc.; II
Eugene Hatke, Purdue University; III	Alan Liby, Oak Ridge National Laboratory/UT-Battelle; III
Russell Healey, Smith Group; II	James Loesch, Johns Hopkins University Applied Physics Laboratory; I
Brett Helm, DPR Construction, Inc.; II	Joe Lonjin, Pennsylvania State Nanofabrication Facility; I
William Hendrick, Sandia National Laboratories: II	Tim Loughran, AdvanceTEC, LLC; I
Donald Henke, HDR Architecture, Inc.; III	Michael Luciani, Gilbane Building Company; III
Julie Higginbotham, Laboratory Design Newsletter; III	William Lynch, Microzone Corporation; I
James Hill, BSA LifeStructures; III	Wade Mayberry, Earl Walls Associates; II
Mark Hoover, NIOSH; III	Scott Mackler, Cleanroom Consulting, LLC; I
Linda Horton, Oak Ridge National Laboratory; III	Ian MacLaren, Kinetic Systems, Inc; II
Louis Hornyak, University of Denver; II	Robert MacGregor, Intelligent Enclosures, Inc.; I
Linda Horton, Oak Ridge National Laboratory; II	John Majewski, Rafael Vinoly Architects; III
Emile Hoskinson, University of California, Berkeley; I	Derrick Mancini, Argonne National Laboratory; III
M. Kamal Hossain, National Physical Laboratory, UK; I	Richard Martorano, Arizona State University; II
Ben Huey, Arizona State University; II	Mike Martindill, Hardin Construction Company; II
Julian Hunt, National Physical Laboratory, UK; I	Sally Mason, Purdue University; III
Robert Hwang, Brookhaven National Laboratory; II	Bill Matsukado, Bechtel—China; II

(continued)

Rod Horning, Technical Manufacturing Corp.; I	Joseph Mattrey, M+W Zander; III
Thomas C. Isabell, JEOL USA, Inc.; II	Theresa Mayer, Penn State University; III
Tim Isle, Shiel Sexton Company; III	Gerard McCabe, Curran McCabe Ravindran Ross, Inc.; III
Claude Jacques, National Research Council Canada—INMS; I	Mike McCann, Pepper Construction Company of Indiana; III
Cheryl Jamison, HDR Architecture, Inc.; I	James McComas, Wilson Architects; I
Mark Jamison, HDR Architecture, Inc.; I, II, III	Timothy Meier, IDC Architects; II
Robert Jamison, Intel Corporation; II	Armand Milazzo, Deutsch Associates; II
David Janes, Purdue University; II, III	Timothy Miller, Purdue University; I
Jodie Johnson, HDR Architecture, Inc.; III	Jim Montgomery, The Cohos Evamy Partners; I
Dave Jones, Air Products & Chemicals, Inc.; II	Julian Montoya, Intel Corporation; II
Joonho Joung, National NanoFab Center; III	Marcel Morabito, St Owe deo Neutyrs; III
David Kaluf, Hunt Construction Group, Inc.; III	Joseph Morgan, Wilson Architects; III
Fida Kanaan, HDR Architecture, Inc.; I	Ralph Morrison; I
Debbie Kenny, Labconco Corp.; III	Jeffrey Roblee, Precitech, Inc.; I
Randy Ket, Hunt Construction Group, Inc.; III	Mark Eliot Rodgers, University of Denver; II
Moyeon Kim, Human & Technologies Corp.; III	Thomas Rogers, Technology 2020; III
Howard Kerr, PCL Construction Management; I	Bob Rose, Georgia Institute of Technology; III
James Murday, Naval Research Laboratory and University of Southern California; I, III	Robert Rose, Georgia Tech—MIRC; II
Gary Nagamori, HDR Architecture, Inc; II	Craig Rossrucker, Abbie Gregg, Inc.; II, III
Jan Nedelka, Eastman Kodak Company; I	Josh Rownd, HDR Architecture, Inc.; III
LuAnn Nelson, Medtronic, Inc.; II	Patrick Ruane, MED Institute, Inc.; III
St Owe deo Neutyrs; III	Josh Rownd, HDR Architecture, Inc; II
Richard Newhouse, Hunt Construction Group, Inc.; I, II, III	Fred Russell, PCL Construction Management.; I
Warren NG, Lawrence Berkeley National Laboratory; II	Timothy Scarlett, Wilson Architects; III
Susanne Nicholls, HDR Architecture, Inc.; II, III	Mike Schaeffer, Brookhaven National Laboratory; III
Scott Nicoll, University of Waterloo; III	Michael Schaeffer, Brookhaven National Laboratory; I, II
George Norek, Argonne National Laboratory; III	Mark Schattenburg, Center for Space Research; I
Robert Novak, Hammel, Green and Abrahamson, Inc.; III	Linda Schmoldt, Armacel; III
David Oh, HDR Architecture, Inc; II	Jeff Schramm, Gilbane Building Company; III
Billy O'Neill, Project Management, Ltd.; I	Paul Schulte, NIOSH; III Samir Srouji, Wilson Associate Architects; II

(continued)

Brendan O'Neill, NMRC; I	Matthew Sears, Sears Gerbo Architecture, II, III
Ernest Orlando, Lawrence Berkeley National Laboratory; II	Jack Seay, The Beck Group; II
Lodovico Osio, SERIN; I	Pete Secor, Shortridge Instruments, Inc.; II
Arthur Otto, CCRD Partners; I	Bea Sennewald, HDR London; I, III
Carlo Pantano, Penn State University; III	Pankaj Sharma, Purdue University; III
Gregory Parker, Currie & Brown, Inc.; II	T.C. Shen, Utah State University; I
Jack Paul, HDR Architecture, Inc.; III	Oscar Horacio Vigna Silva, Natl. Synch. Light Laboratory; I
Robert Peale, University of Central Florida; III	Neal Shinn, Sandia National Laboratories; II, III
Adrienne Perves, CEA-MINATEC; III	Jim Short, Wafertech; III
Stephen Phillips, Stantec Architecture; III	John Sidarous, Argonne National Laboratory; III
Gianluca Piazza, University of Pennsylvania; III	Chris Skiba, Purdue University; III
Diana Prideaux-Brune, University of Massachusetts Lowell; III	Kevin Sladovnik, HDR Architecture, Inc; II, III
Tura Patterson, Affiliated Engineers, Inc.; II	David Slattery, Smith Group; II
Gerald Paulus, City of Mesa, Arizona; II	Rob Smalley, Mechanical & Engineering Contractors, Inc., Arizona State University; II
Mathew Perkins, HDR Architecture, Inc; II	Charlie Smith, Pepper Construction Company of Indiana; III
Pat Peters, DL Withers Construction; II	Michael Smyser, HDR Architecture, Inc; II
Tina Prestridge, Georgia Tech—MIRC; II	Gregory Snider, University of Notre Dame; III
Doug Perry, HDR Architecture, Inc.; I	Todd Snouffer, NIST; II, III
Francisco Ramos, Northwestern University; II	Elaine Solomon, HDR Architecture, Inc.; III
Gregory Raupp, Arizona State University; II	Michael Somin, Earl Walls Associates; I
Manoj Ravindran, Curran McCabe Ravindran Ross, Inc.; III	Ahmad Soueid, HDR Architecture, Inc.; I, II, III
Chris Rayner, Pepper Construction Company of Indiana; III	Gary Spinner, Georgia Institute of Technology; III
Mark Reed, Wilson Associate Architects; I, II	Samir Srovji, Wilson Architects; I
Ron Reifenberger, Purdue University; I, III	Jack Stellern, Oak Ridge National Laboratory; I, II, III
Wellington Reiter, Arizona State University College of Architecture; II	Mark Stephens, EPRI PEAC Corporation; I
Jon Repp, Engineering Technical Services, Arizona State University; II	Susan Stewart, HDR Architecture, Inc.; I
Donald Rerko, URS Corporation; III	Roger Stewart, HDR Architecture, Inc; I
Pat Richardson, Oak Ridge National Laboratory; III	Keri Stipp, Pepper Construction Company of Indiana; III
Steve Riojas, HDR Architecture, Inc.; I, II, III	Glenn Stowkowy, Stantec Consulting, Ltd.
Michael Roanhaus, HDR Architecture, Inc.; I	Chris Stolzer, Kiewit; II
Steven Robinson, URS Corporation; III	Joseph Stroscio, NIST; I

(continued)

James Robinson, The Austin Company; II	Lewie Wallace, Purdue University Construction Dept.; I, III
Mark Strnad, Abbie Gregg, Inc; II	Jim Walters, Purdue University Construction Dept.; I
Tim Studt, R & D Magazine; III	John Weaver, Delphi Delco Electronics Systems & Purdue University; I, III
Don Surina, Sundt Construction, Inc; II	John Weinman, Perkins & Will; III
Clare Swanson, HDR Architecture, Inc.; III	Steve Westfall, Tradeline, Inc.; II
E. Clayton Teague, NIST & National Nano-technology Coordination Office; I, II, III	Joseph F. Wheeler, Sundt Construction, Inc.; II
Richard Torres, ASML, US, Inc.; II	Ira Winston, University of Pennsylvania; III
Norm Toussaint, HDR Architecture, Inc; II	Mark Winter, Abbie Gregg, Inc.; II
Stephen Treado, NIST; I	James Whetstone, NIST; I
Mark Tuominen, University of Massachusetts Amherst; III	Robert Wilson, Clark Ventures; I
Daniel Ugarte, National Synchrotron Light Laboratory, Brazil; I, III	Ferran Xinxo, CNM-CSIC; I
Eric Ungar, Acentech, Inc.; I	Hui May Yang, HDR Architecture, Inc; III
Jorge Urrutia, NIST; I	Amir Yazdanniyaz, ARUP; II
Brent Vogles, The Whiting-Turner Contracting Co; III	Don Yeaman, M+W Zander; III
Kelly Vincent, HDR Architecture, Inc; II	Lara York, Abbie Gregg, Inc.; II
Lou Vitale, VitaTech Engineering; I, II, III	Frederic Zenhausern, Center for Applied Nanobioscience, Arizona State University; II
Andras Vladar, NIST; I	Rolf Ziemann, Northwestern University; I
Tom Vogt, USC NanoCenter; III	Ted Zsirai, HDR Architecture, Inc.; I

Affiliations are as of the time of the workshops.

Index

© Springer International Publishing Switzerland 2015
A. Soueid et al. (eds.), *Buildings for Advanced Technology*, Science Policy Reports,
DOI 10.1007/978-3-319-24892-9

Printed in the United States
By Bookmasters